Genetic Priorities for Conservation of Native Honey Bees

Dorian Pritchard

Northern Bee Books

Genetic Priorities for Conservation of Native Honey Bees

Published in the United Kingdom by
Northern Bee Books,
Scout Bottom Farm,
Mytholmroyd,
West Yorkshire HX7 5JS
Tel: 01422 882751
Fax: 01422 886157

www.northernbeebooks.co.uk

ISBN 978-1-914934-89-6

Design and artwork, DM Design and Print

Cover illustration: The author's native Northumbrian bees.

Illustrations by the author

Genetic Priorities for Conservation of Native Honey Bees

Fig. 1 One of the author's native worker bees on Clematis.

DEDICATION

To my teachers, students and the bees, for their interest, inspiration, and continuing support

CONTENTS

FOREWORD

Like Dorian, I started beekeeping in the 1970s. I joined BIBBA (which then stood for the British Isles Bee Breeders' Association) and became a committee member. I was privileged to be asked to compile the book *The Honeybees of the British Isles* from the articles and papers left by BIBBA's Director Beowulf Cooper after his untimely death in 1982.

After a scientific training Beowulf had worked for many years as an entomologist for the Government's Agricultural Advisory Service, as well as being an experienced practical beekeeper. In his book he helped to debunk the commonly held belief, promoted by Brother Adam and others, that the native honeybee *Apis mellifera mellifera* had been wiped out in the British Isles by the 'Isle of Wight Disease'. He described its genetic characteristics, both morphological, being an early practitioner of wing morphometry, and behavioural, and discussed its management.

Through BIBBA and also SICAMM (the European dark bee association) I became acquainted with Dorian Pritchard, then Lecturer in Human Genetics at the University of Newcastle-upon-Tyne and later co-author of the standard work *Medical Genetics at a Glance*.

I find the present book a worthy successor and complement to Beowulf's, written like his by a scientist (this time a highly qualified geneticist) and experienced and successful practical beekeeper.

As Dorian shows, since Beowulf's day the study of honey bee genetics by wing morphometry and DNA analysis has advanced considerably. Meanwhile, many commercial beekeepers and queen breeders, in their quest for ever greater honey production, have taken the uniformity of their bees to even greater lengths, importing bees and/or queens from distant areas, and also routinely treating prophylactically for varroa and other diseases.

At the other extreme, conservationist proponents of 'let-alone', 'natural' or 'Darwinian' beekeeping have also become more evident. Dorian's genetic expertise and his many years of practical beekeeping enable him to steer a sustainable and yet productive course within this field, with scientific authority. He comes down firmly on the side of keeping only local bees which are adapted to the prevailing climate and environment, while conserving natural variation, and recommends beekeepers "to allow *all* native colonies to reproduce without restraint, allowing nature to make her own choice of survivors." Compatible management techniques can aid this process and still provide a worthwhile honey surplus.

Philip Denwood

Main Board member, **SICAMM**

Former Committee member and magazine editor, **BIBBA**

ACKNOWLEDGEMENTS

My thanks go first to Jeremy Burbidge for his kind invitation to write this book. Also, to Philip Denwood for his wise judgement and strong support over many years, for reading the manuscript and for his generous foreword. Ian Robinson has been a most supportive friend, with his erudite and lucid criticism of my ideas and I thank him also for his help and support in many other ways. Kyle Miller I thank for valuable discussions on our trips to the hills, interspersed with hilarious stories that kept us laughing all day; also for use of his photograph of a native queen and workers in Fig. 9. I remember the late John Dews for his friendship, his wise counsel and valuable exchange of ideas and thank him for his permission to use his excellent photo of a native worker on plum blossom in Figure 2.

I am also most grateful to my son Hamish for valuable discussions on glaciation and glacial melting and especially for his expert summary of the climatic issues surrounding the potential demise of the Gulf Stream. The late John Theobalds was one of the best beekeepers I have known and I remember him gratefully for sharing his insights into native honey bee management. Also, the late Amy Nicholson for many tips on how to work with bees and the late Rev. Eric Milner for teaching me how to do wing morphometry. I am also grateful to Martyn Farrer for his friendship and generous support of my sometimes-way-out ideas and the members of Hexham BKA for their supportive interest.

It was my grandmother, Nanna, who sowed the seeds of my interest in bees, but it was really fired by an anonymous Pembrokeshire beekeeper who astonished me as a boy by showing me a queen and attendants who had just arrived by post in a matchbox, and a heavy shelf of clear, red honey that had all come from one hive! Thank you for that inspiration!

.

Lastly I thank my wife Penny for her tolerance of multitudes of dangerous insects in the garden and sometimes the house, her fearlessly catching the occasional swarm and for putting up with sticky doorhandles, sticky patches on the lino and my frequent cluttering up of horizontal surfaces with books and typescript.

AUTHOR'S FOREWORD

The principles of native bee keeping are not the same as those of "natural beekeeping", although there is much overlap between the two. My approach is to keep bees in as close as practicably possible to a natural situation, but to run them in standard beehives. I use British Nationals with standard boxes and frames. Smith hives, devised by Willie Smith of Innerleithen in the Scottish Borders are also popular in my area of Northumberland, but I prefer Nationals, largely because the hand and finger grips are better.

During the Great Ice Age, the ancestors of *A. m. mellifera* survived in refugia in Spain and the South of France. They came to Britain as the glaciers melted, crossing the Land Bridge from mainland Europe before it was swept away 9000 years ago. Descended from cold-adapted ancestors, they grew to revel in the returning warmth, but endured the unexpected chill of the Younger Dryas, that helped exterminate the Pleistocene megafauna, and Mountain Aven (*Dryas octopetala*) returned for 1200 hundred years to our British lowlands. Honey bees endured the cold blasts that drove our Iron Age forebears from their hill forts and long-term blackouts caused by volcanoes in El Salvador, when dust blotted out the sun and 80% of the human population of Scandinavia starved to death. AD 536 was described as "the worst year in history", with 24-hour darkness for 18 months and summer temperatures plunging to 2°C. They came through all that! The pleasant Medieval Warm Period began in AD 950, but was brought to a chilly end by the Little Ace Age in 1305, ending as recently as 1860 and including the Great Frost of 1709, with temperatures down to -12°C and "frost fairs" held on the frozen river Thames. Our bees have faced hurricane winds, deluges, droughts, badgers, bears, woodpeckers, ants, honey buzzards, great tits, hornets, wasps, wood mice and shrews, starvation, foul brood, nosema, acarine disease and yes, even Isle of Wight Disease, to say nothing of the host of pests and diseases that destroyed their forage! The resultant survivor was virtually indestructible. If it were not, it would have gone long ago, along with all the other failures. In the British climate it was the ultimate survivor and our greatest national treasure!

But look at the present situation. In 2014, COLOSS published a survey which showed that, although local stocks everywhere in Europe can out-survive exotic imports, almost all of those locals now cannot live for even **THREE YEARS** without the support of medication! (see Büchler et al, 2016 and Chapter 4.) After 50 million years of healthy independence, 30 years of human mishandling have all but done for them. But the COLOSS survey did not include Britain, and I think our bees can show the rest of Europe a thing or two!

On their own turf, all races of honey bee were, until about 100 years ago, supreme survivors. They were the unadulterated sons and daughters of the best of the best. Transfer them to a different environment and that is no longer the case, which is the situation for all exotic imports into Britain. But, on their own ground our own natives are still the masters, capable of surviving every climatic insult imaginable, provided their genetic base has not been compromised by matings with foreign bees, and provided they are given just a few basic necessities.

These are a secure, dry, draught- and water-proof cavity of about 40 litres capacity, with a small entrance preferably facing the 10 o'clock sun, good forage through most of the season and no interference from predators, including humans. Their main requirements are for warmth, food largely in the form of nectar, as a pleasureable source of carbohydrate, and pollen, providing an abundant supply of protein and lipid. They usually acquire trace elements, essential amino- and fatty acids, vitamins, etc. and beneficial live gut bacteria as minor components of nectar and pollen. White sugar, sucrose, is frequently fed to bees as a substitute for the honey their keepers have taken from them. It should be noted that white sugar is the purest chemical we humans eat, so, apart from supplying one molecular species of carbohydrate which is also pleasantly sweet to them as well, it is essentially junk food even for bees.

If warmth and food are not immediately available to bees, the answer would seem to be for the beekeeper to supply both artificially - and this is what all the books tell you to do. I say this is **NOT** the response you should adopt, if your aim is to maintain native bees in good health and good heart. This is because that approach undermines the natural selective forces that, over countless generations have given them the attributes you and others admire: their strength, endurance and self-sufficiency. If you continue to give them generous material support (as the books tell you to do), your stocks will eventually lose those strengths. They will end up as dependent, domestic wimps, having always fought their place like Spartan warriors!

Your aim should not be to supply the bees with their obvious requirements directly, but instead to put them in positions where they can satisfy those needs by their own efforts. So, you should not give them sugar, or any other food, except in emergency, instead you take them to sites where, with a bit of effort, they can fill their storage cells with the nectars and pollens to which they and their kind have become accustomed for the past 10,000 years and more. This restores their self-confidence, keeps them in good heart, and reinforces the role of the genome that makes them what they are.

Likewise, you should arrange their hive interiors so they can keep themselves warm by ways they discovered by trial-and-error way back in the Great Ice Age. You don't insulate their hives, instead you allow their hives and the bees they contain to get cold as environmental temperatures fall, so they are forced to revisit those ancient methods of their ancestors. As has been shown over and over again, for native bees, the best policy is NOT to insulate their hives for winter, but deliberately to keep them cold. If you do so and they survive (as is usually the case), their survivalist genes are brought to the fore, and ancestral behaviours reinstated in their lifestyle.

If they have forgotten their ancient traditions, in natural circumstances they would die, but in the apiary situation, competent beekeepers can identify such under-performers and replace their queens with others whose genomes still sparkle with the priceless genetic jewels of distant ancestors.

So, the principle is: provide your bees with the means to achieve success, not the actual objects of that success. It's like teaching a hungry man to fish, rather than supplying him with fish someone else has already caught. Don't feed them with sugar because you consider the forage too poor, take them to better forage. Don't insulate a hive that is adequate in summer, for their use in winter, keep them cold, but give them space to form and operate a winter cluster. Don't remove their need to struggle, encourage them to take on their problems - and conquer them!

The influences that created native bees in the first place are the forces of natural selection that are still with us. We do the bees a disfavour if we insulate them from those forces – they **NEED** to fight starvation and cold! They delight in that battle, just as we do, and they come out of the battle reinvigorated for the season to come. Brother Adam carried out many trials on the merits or otherwise of insulation and he found repeatedly that post-winter colonies at first appeared stronger when their hives had been insulated, but were rapidly overtaken by the vigorous occupants of un-insulated ones.

If necessary you should even exaggerate the ancient and traditional natural selective forces and allow your charges to indulge themselves in fighting for survival; remove the need to fight and you'll end up with wimps. Ask yourself: do you manage your bees like old folks in a care home, or like trainees for the SAS? If the former, you can congratulate yourself on being a nice person, but if you want healthy, self-confident and happy bees you should be doing the latter.

Native bees are psychologically, physiologically and behaviourally equipped to respond to existential challenges; if they weren't they would have died out long ago. They do so with vigour and display confidence when they have defeated those challenges. It is like our eagerness for adventure sports and they, like us, benefit emotionally from seeing off a threat. But the style of management we are everywhere told to adopt with bees is to remove those threats and destroy the challenges that have always kept survivors strong and excited about life.

You will know the saying: "*When the going gets tough, the tough get going*", well that applies to bees as well. We should cherish and harbour those bees and those genes that make it so.

Since the glaciers of the Great Ice Age dwindled, pioneer families of Dark honey bees spread north through what became England and were brought to a halt before the might of Cheviot. There in Northumberland they established communities of self-sufficiency and there they exist to this day. Those are the bees I have known, trust and come to love.

Officially, the native honeybee is extinct, or never came here under its own wing power, but was instead brought in by the Romans. But that is not the case, their DNA tells a different story, and the archaeologists can point to relics that long preceded the tramp of Caesar's legions.

But the Roman legacy gives us the first record of beekeeping in Britain, in an erotic statuette of Priapus dug up by one of my beekeeping students that would have been placed on guard beside the apiary, or perhaps to improve its queens' fertility. There is also a handwritten order on a sliver of birch wood, for "*duo lini mellari*", two honey cloths. Both were dug up at Vindolanda, a Roman officer holiday retreat just on the sunny side of Hadrian's Wall, left behind when the Romans returned home in 410 AD. This is near where I live and have my home apiary and where I am writing this book.

From the Northumbrian perspective, internationally the honey bee situation looks in a mess and the roulette wheel of history appears weighted toward its decline, perhaps extinction in mainland Europe and North America.

We do not share the problems we see reported from much of the world and this text is meant as a lifeline we're throwing to help those who seem to be going under, in the hope they can still be saved. For those who need it, good luck to you all!

Dorian Pritchard B.Sc., Dip. Gen., PhD, MRSB.

"All over England, from Northumbrian coasts,
To the wild sea-pink blown on Devon rocks,
Over the merry southern gardens, over
The grey- green bean- fields, round the Kentish oasts,
Through the frilled spires of cottage hollyhocks,
Go the big brown fat bees, and wander in
Where dusty spears of sunlight cleave the barn,
And seek the sun again, and storm the whin,
And in the warm meridian solitude
Hum in the heather round the moorland tarn."

From *"Bee-Master"* by Vita Sackville-West, 1892-1962

Chapter 1

BASIC CONSIDERATIONS

My background and orientation

I began beekeeping at the end of the 1970s when I had just taken up the post of Lecturer in Human Genetics at the University Medical School in Newcastle-upon-Tyne. Almost everything I knew with regard to my profession I had learned from books, so for a change, I decided I would learn beekeeping from people and from the bees themselves. So I hardly opened a handbook on practical beekeeping until 20 years later, when I was asked to teach a beginners' course at Kirkley Hall Agricultural College, Ponteland.

What I read shocked me, as the directives they and the popular beekeeping magazines gave were distinctly different from what I had learned from local beekeepers. I came to realise that, with the possible exception of Robert Couston's, *Principles of Practical Beekeeping*, these books were written for beekeepers in the South of England, trying to maintain bees that originated around the Mediterranean, in a part of the world for which they are not adapted and where they shouldn't be anyway! For example, they told me I should insulate my hives in winter, but not in summer and not take them to the heather, as that would give them dysentery, but feed them instead with toxic quantities of white sugar. Or treat them with chemicals for diseases we never see.

I realised also that most of the Bee Diseases Inspectors had acquired their knowledge from some higher power, rather than by watching the bees themselves and this was especially so when varroa arrived! My native and near-native bees turned out to be resistant to varroa, though everyone who didn't know me, or recognise I have been a research biologist for half a century, said I must be mistaken. And that indeed is still the case.

Moreover, as a biologist, I knew that native species were created and kept healthy by the forces of natural selection, in accordance with Charles Darwin 's "*Theory of Evolution through Natural Selection*", but the experts who wrote the books and articles about beekeeping were telling me I should shield my bees from natural selection in every emergency!

The priorities for conservation and those for their commercial exploitation are a world apart and the beekeeping "authorities" it seems are only versed in the latter. This text, though, will take the alternative course.

This book is therefore intended as a practical guide to the long-term conservation of our native honey bee, *Apis mellifera mellifera* L. It derives from lectures I gave to SICAMM and BIBBA conferences in the summer of 2023. Those were presented with the intention of alerting honey bee conservationists to the dire situation in mainland Europe, the United States and elsewhere. It draws attention to what I believe are uninformed management practices that are placing honey bees increasingly in peril, including queen propagation methods that wantonly destroy the natural genetic variation essential for their survival.

As a reference, I am stating here that I manage some 30-40 timber Modified British National hives, plus a few of dense, expanded polystyrene, at six apiary sites. My bees have always been of local origin with *A. m. mellifera* wing morphometry, but occasional, probably *carnica*, contamination.

A bit over 20 years ago mitochondrial DNA analysis revealed them to include several queen lines and recently SNP (single nucleotide polymorphism) analysis of their chromosomal DNA showed North European M-lineage content of up to 94%. They are gentle and honey yields are good, but not outstanding. I have not treated against Varroa since 2002; their resistant behaviour includes aggressive grooming (see Appendix), but there is no evidence of the uncapping and sometimes recapping, known as Varroa Sensitive Hygiene, or VSH.

In accordance with British parlance among beekeepers, the term "hybrid" here refers to offspring of crosses between subspecies, which are fully fertile, rather than between species, that are not.

Keep local and keep close to nature

For setting up and managing a conservation apiary, you need to use local bees and to keep their forage also of a locally traditional type. Those bees evolved with that forage; no other bees are likely to be better adapted for long-term survival and it could be that no other forage is as suitable for their long-term welfare.

Do not import foreign bees to conservation areas. Identify local stocks that both thrive and show the external features of the race you are trying to conserve. In the case of British natives these include a generally dark appearance, narrow tomenta (the pale, transverse bands on the abdomen) and no coloured bands. Their dorsal abdominal body "hair" is longer than the breadth of the broadest section of their hind legs and their tongues are relatively short.

Fig.2. The native British "Dark" honey bee, *Apis mellifera mellifera* L. on plum blossom.
(Photo by John Dews, reprinted with permission.)

Work *with* natural selection, not against it.

Advice to beekeepers from "the experts" frequently involves combating some aspect of natural selection in the interests of an easier life, or more profit for the beekeeper. But it is to natural selection that we owe the creation and survival of all our native species, through their continual adaptation and survival in the face of adversity over tens of thousands of years. The discrepancy arises because the goals of commercial exploitation are short-term and self-oriented and only rarely address long-term considerations like ensuring honey bee survival for another 10,000 years.

Beekeepers are directed to thwart natural selection because it reduces profits and yet those advisers maintain the expectation that nature will ensure honey bees' health and continued survival despite their inroads. Hive economy is vital to ourselves as well as the bees and yet I have found very little literature that tells us how to manage it.

The main difference between commercial bee breeding and breeding for conservation is the amount of genetic diversity aimed for. Breeding of commercial strains focusses on selection of just a small aspect of the whole of the genetic diversity of a given population, whereas conservation breeding essentially aims to retain it all, regardless of commercial considerations. Ideally this would involve very many colonies and as large a population as possible, to maximise retention of diversity.

Conservation has a particular aim of ensuring survival of keystone species and the honey bee is one of these, but in addition the lesser species upon which our natural ecosystems rely must also be preserved in good health, while ensuring the good health and wellbeing of our own species also.

One way we can work with natural selection is by exaggeration of seasonal temperature extremes. This we can do by *insulating our hives in summer and withdrawing that insulation in winter*. This however, is counter-intuitive for most beekeepers and one of the most difficult ideas to accept. Another is the non-provision of supplementary feeding, and in its place, transfer of hungry colonies to better natural forage. Non-medication of varroa infestations is another tactic that encourages bees' natural independence.

Natural selection keeps species healthy by eliminating weaklings. In the wild it is the main agent of "improvement", a word that has a very different meaning in the context of commercial beekeeping. This keeps the average fitness of current generations high and also improves the prospects for future generations, by removing the genetically unfit. Natural selection is therefore a vital influence and you should allow it to take place! So, do not pamper native bees, they specialise in tough conditions. If unable to cope, unite them under a stronger queen, or requeen them from more competent stock.

However, colonies need time to build up their strength. I usually consider a normally developing colony with 5 frames of brood in a timber National hive to be the equivalent of a boy of about 15, not quite able to look after himself, but well on the way. At 8 frames, it is like perhaps an 18-year-old and should be able to cope without much help on my part. In anticipation of winter I assess my colonies in August and, if they are on 5 or fewer frames of worker brood I make sure they have one or two deep frames of stores, or feed them well and give them top-insulation for the winter.

If the colony has 8 frames of brood I make sure the stores are good, and do not insulate for winter. However, in Spring I may give any slow colony a half pack of Candipolline Gold (a pollen based and nutrient supplemented food) and add top insulation when I see pollen being taken in, which tells me there is a laying queen and brood is being raised. I then leave the insulation on all summer to help them draw the comb and raise strong brood.

The percent viability of worker brood is particularly important, as well as the number of frames of worker brood and this is dealt with later.

CHAPTER 2

DISCRIMINATION OF HONEY BEE SUBSPECIES

Wing morphometry

The easiest way to gain some idea of how close to native your bees are, is by wing morphometry. Around 30 bees of the same colony must be sacrificed, and the most humane and convenient procedure is to catch the live bees in a glass jar at the hive entrance, cover it with muslin and place it in a deep-freeze (not the freezer compartment of a refrigerator) for 48 hours. Then take off their right wings and mount these on glass slides, using sugar syrup at their bases to stick them to the glass. Each slide is then projected onto a white screen and measurements are made of the positions and lengths of several wing veins. This is normally facilitated with paper templates that help the operator discern two measures known as the Discoidal Shift (DS) and the Cubital Index (CI) (see Fig. 3).

The Cubital Index is the ratio of the length of line BC (i.e. "a") to that of line AB (i.e. "b"). The ratio is "a" divided by "b" and for native *A. m. mellifera* should be just below 2.1. Values for historical British museum specimens were quoted by Ruttner, Milner and Dews (1990) as 1.40 to 2.09 (mean: 1.79).

The DS is revealed as the displacement of point D with respect to a vertical line dropped through vein junction H and is measured in degrees to the left, termed "a negative shift" or "positive" if the shift is to the right. Values for the British museum specimens were in the range: +1.0 to -2.87 (Ruttner, Milner and Dews, 1990).

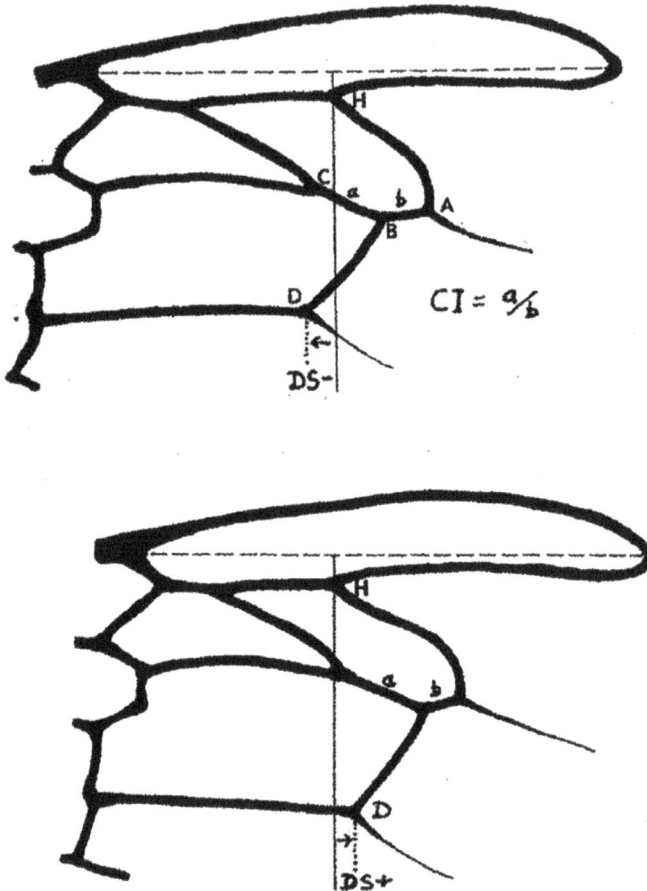

Fig. 3. Wing morphometry. Native *A. m. mellifera* show a mean negative Discoidal Shift (i.e. point D is to the left of the vertical line) as in the upper picture and a mean Cubital Index (a/b) of less than 2.1. The lower picture shows a wing with a positive DS as in *A. m. ligustica* and *A.m. carnica*.

Those two values for each right wing are then plotted on a graph as in Fig. 4, which shows the scan for a wild Northumberland swarm led by a queen we named Ella. The mean value of the 30 readings has a DS of -4 and a CI of 1.6, which are well within the defined values that specify *A. m. mellifera*, the Northern Dark Bee Fig. 5.

There have been several contributors to wing morphometry, but the use of optical projection, quantification of the degree of shift and the 2-dimenional graph were introduced by John Dews, in whose memory I term the DS/CI graph a " Dews' plot".

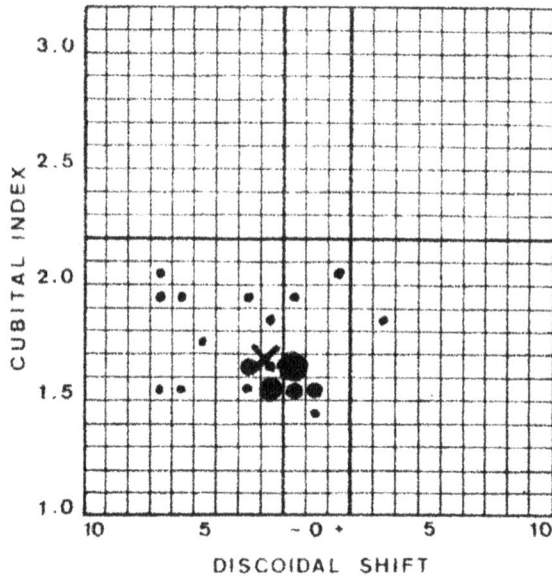

Fig. 4. A Dews' plot of the colony of Ella, queen of a wild Northumberland swarm. The cross represents the mean.

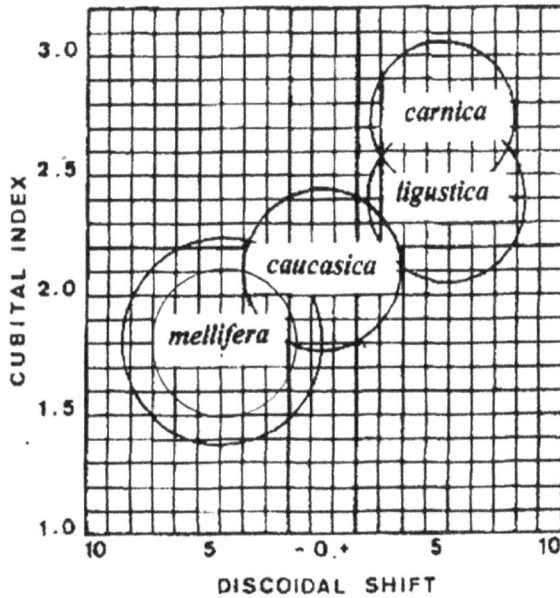

Fig. 5 shows the approximate positions of the equivalent readings for the most likely contaminants of *A. m. mellifera* in the UK, *A. m. carnica*, *A. m. caucasica* and *A. m. ligustica*.

Figure 6 shows the values for largely unrelated *A. m. mellifera* stocks at eleven locations within a radius of 50 miles. The overall mean values are Cubital Index, 1.74, Discoidal Shift, -2.7." (i.e. negative 2.7)

A note of caution is necessary here. In recent years morphometric analysis has been computerised and one of the selling points for the computer programmes has been the claim that it can tell you the "percentage mellifera" of your sample. This is done by expressing the number of wings that fall within the racial boundaries, as a percentage of the total number in the sample. This, however, is a bogus claim and for two reasons. The first is that the boundaries for racial classification were delineated by Friedrich Ruttner as defining the limits of *mean* values, not those for individual bees. The second is that there are many reasons why readings are scattered. All the bees in a colony that are descended from one queen are a complex expression of the genomes of both parents and it is unlikely to be simply that some bees have a markedly greater percentage of the genome of one race rather than another, unless the drone population is markedly diverse.

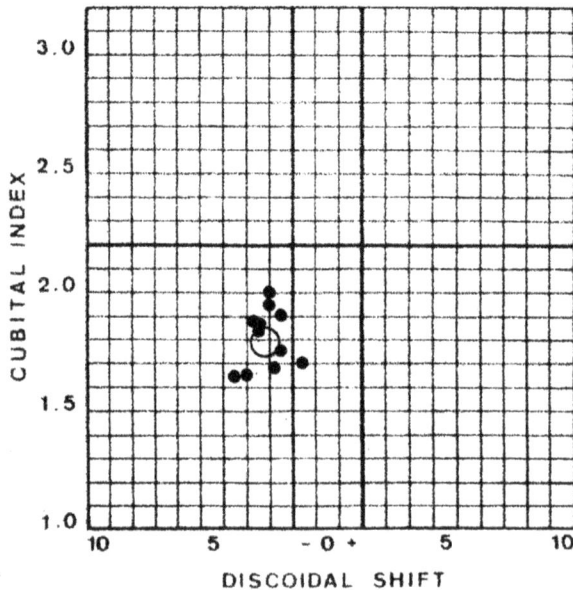

Fig. 6. The combined Dews' plots for eleven Northumberland stocks, showing all mean values in the lower left quadrant, characteristic of *A. m. mellifera*. The open circle represents the overall mean at CI:1.74; DS: -2.7.

DNA analysis

DNA analysis has proved invaluable in investigations into racial identity and evolution, but the methods employed are constantly changing. Around the turn of the millennium sequencing of a small, hypervariable section of the mitochondrial DNA was performed on many stocks and demonstrated that British queen lines had characteristics that are unique to the British Isles. One of my own stocks derived from a swarm caught in north Northumberland turned out to be of the same queen line as those that had been introduced to the Inner Hebridean island of Colonsay (Bo Vest Pederson, personal communication). Those bees are now recognised by the Scottish Government as true Scottish natives and protected by law. Both English and Irish stocks have characteristic markers and unique sequences.

Since that time, attention has been directed at the nuclear, chromosomal DNA and other methods employed. Of these the currently most favoured involves identification of a large number of Single Nucleotide Polymorphisms, or SNPs, pronounced "snips". These are the equivalent of alternative spellings of the same word, like Jon and John, or Grey and Gray. The first could represent an insertion or deletion of "h", the second a substitution of "e" for "a", or *vice versa*. The great strength of this method is that it covers virtually every part of every chromosome and once the linkage relations of important genes to specific SNP sites is known, it permits selection of stocks at the molecular level for breeding

Fig.7 shows a report on the analysis of one of our Northumberland stocks by this method, carried out by the Edinburgh firm Beebytes. The lower array shows the individual findings for each of the SNP loci examined, black columns representing an M-lineage marker at that site, yellow columns those of C-lineage. The M-lineage is the one of five or more ancestral lineages that evolved north of the Alps and contains *A. m. mellifera*, the C-lineage, its greatest present competitor, evolved south of the Alps and notably includes *A. m. carnica* and *A. m. ligustica*.

The arc indicates the total percentages of M and C markers in that colony.

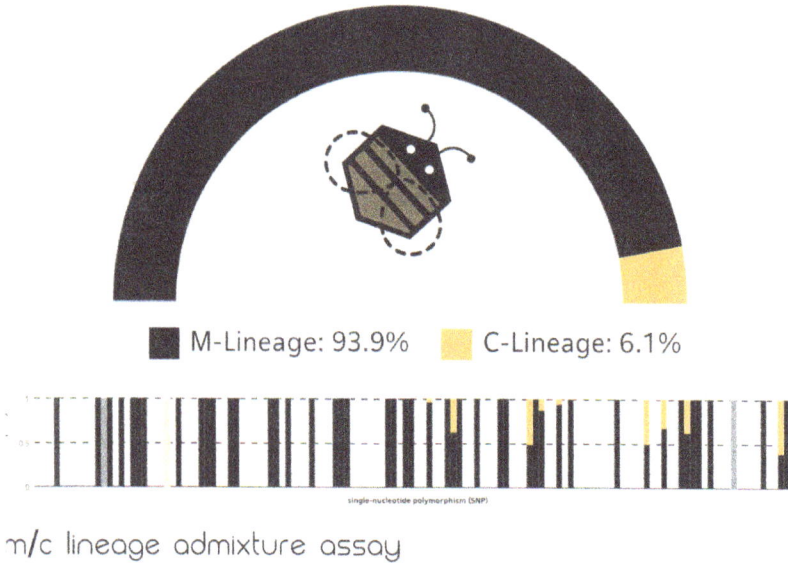

M-Lineage: 93.9%   C-Lineage: 6.1%

single-nucleotide polymorphism (SNP)

m/c lineage admixture assay

Fig. 7. An SNP analysis of the DNA of a Northumberland native colony headed by the grand-daughter of "Ella". It reveals the proportion of M lineage DNA to be 93.9% in this colony. (Analysis courtesy of Beebytes.)

The importance of genetic variation

Nils Drivdal and Josef Stark, co-founders of SICAMM (*Societas Internationalis pro Conservatione Apis Melliferae Melliferae*), holds or held that the secret weapon responsible for native species' success is their genetic variation. This is conspicuously absent from commercial strains and may explain their disturbing vulnerability (see below). A rule in the conservation of all species and subspecies is to preserve their natural variation and not do as many bee breeders do, which is to aim for uniformity of appearance and uniform behaviour.

When disease or other hazards hit a population, if there is no genetic difference between its members, all are likely to succumb or survive together, whereas if the population has not been bred for uniformity it is likely some will survive even if others succumb. And having been through that selection, the population later will be better able to fight off similar hazards.

So, do not aim for uniformity of the stock you are trying to conserve.

Preserve queen lines

A queen line is the theoretical line connecting a present-day queen to her ancestors. Mitochondria are present not in cell nuclei, where the chromosomes are, but in the cytoplasm of virtually all bee (and human) body cells. These are passed on to the next generation in the eggs, or ova, but sperm are too small to carry them, so they are not passed on by males. However, the mitochondria in body cells and ova carry their own DNA and sequence variants in this act a bit like our surnames, indicating the origins and affinities of present-day individuals, but linked through maternal rather than their paternal relatives.

Queen-line-specific mitochondrial DNA variants enable us to trace our bees' ancestry, as shown in Fig. 8. The latest deduction is that *A. m. mellifera* is the common ancestor of the other European, African and Near-Eastern races, not a later subspecies that came "out of Africa", as has previously been suggested.

Queen lines vary in their vigour and prolificacy, so it is natural that the most active will become predominant over others. ***Therefore we must oncentrate on raising new queens and drones from all queen lines, to help conserve their original natural range of genetic variation.***

An especially important rule is: ***do not propagate multiple daughter queens from single breeder queens, to the exclusion of others. We must retain and maintain all queen lines and also aim to mate our virgin queens with drones from unrelated queen lines.***

Outbreeding can be encouraged by inserting drone comb or drone foundation in colonies of unrelated queen lines, or by temporarily relocating virgin queens to more distant apiaries for mating, then returning them to their home apiaries. Insertion of one comb of drone foundation in an 11-frame brood box has no appreciable effect on honey yield.

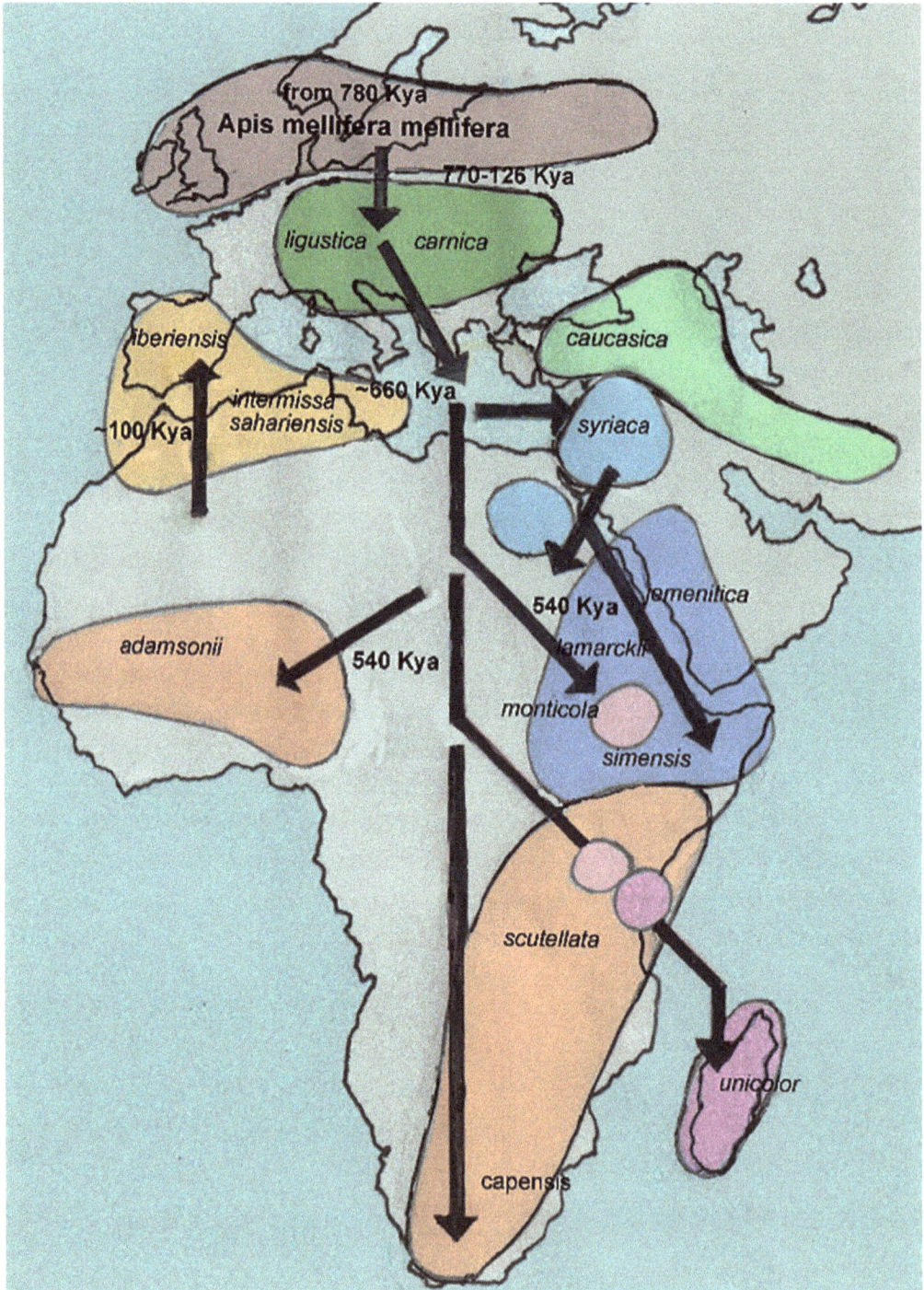

Fig. 8. The pathway of evolution of honeybee mitochondrial DNA. (Redrawn from Carr, S.M. 2023 www.Nature.com/scientificreports) Kya= Kilo-years ago

Chapter 3

THE BRITISH "DARK" BEE

The Native bee in Britain

The great pioneer in native honey bee conservation was Beowulf Cooper, an entomologist who carefully characterised the relict populations of native British bees. Fig. 9 shows a native "Dark" queen marked red among native workers, her daughters.

Fig. 9. A native *Apis mellifera mellifera* (*A .m .m.*) colony showing a laying queen and workers. (Photo: Kyle Miller, with permission.)

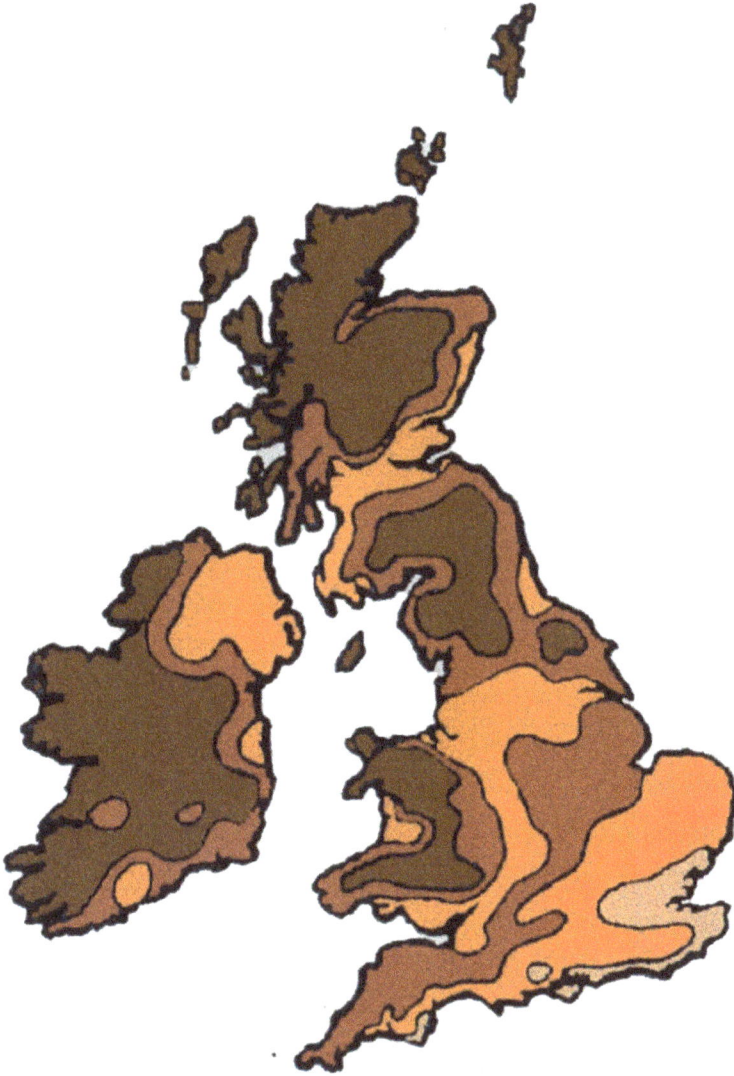

Fig. 10. Degrees of "nativeness" of honey bees as indicated by wing venation studies in the late 1960's. (Redrawn from Cooper, 1986*).

The first map (Fig. 10) is redrawn from one in Cooper's book *"Honeybees of the British Isles"*. It shows areas of high (dark) to low (light) degrees of what Cooper called "nativeness", based on wing morphometry readings. We believe the native heartlands are essentially the same today, 2024.

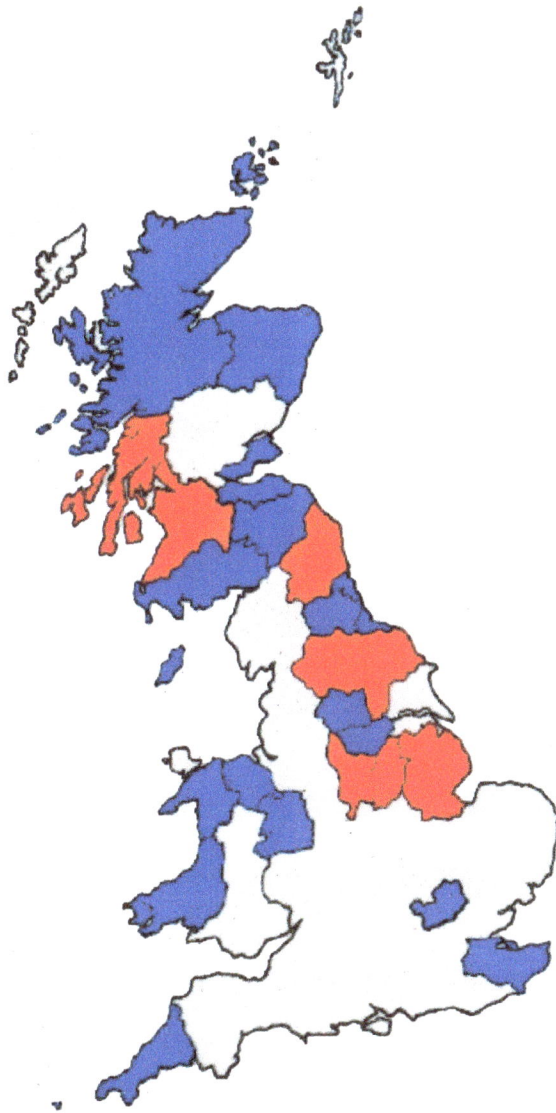

Fig. 11. The known recent distribution of *A.m.m.* in mainland Britain in 2004.

The second map (Fig. 11) is based on reports by BIBBA (Bee Improvement and Bee Breeders' Association) breeding group leaders, of counties where wing morphometry had shown the existence of *A. m. mellifera* colonies, shown in blue, and those, shown in red, where the morphometry had been followed up by positive (native) mitochondrial DNA reports.

Obtaining native bees

I have acquired native swarms either by enticing them into bait hives, or by housing free swarms. I have purchased good examples of natives from other beekeepers and re-created colonies indistinguishable from natives (by the methods available at the time), by "back-breeding" from "near-natives".

That a colony has swarmed indicates its competence to thrive in that area and collecting unrelated native, or near-native swarms strengthens your stock, but the usual concerns of introduction of disease and bad temper should always be considered. Swarms from distant sites may not do well in your area, so I suggest a radius of 50 miles as a practical upper limit to the acceptable range for adoption.

My starting point for back-breeding was a near-native nuc acquired from an elderly local beekeeper. "Back-breeding" involved maintenance of its line in very severe conditions and identifying the strongest survivors at spring inspections. I have only rarely used wing morphometry to guide my selection, as this could result in creation of mimics with the appearance of natives, but lacking their special properties. On one occasion I did use it to help me choose between three stocks and bred only from the chosen one (see Fig.12).

One character unique to native *A.m.m.* is the habit of storing pollen among the brood and between the brood and the frame bottom bar. This was something I had not selected for, but it appeared spontaneously in several hives in the same year and I considered that a strong endorsement of my selection method (see Fig. 13). Overall appearance, wing morphometry and analysis of mitochondrial DNA failed to distinguish descendants of the best of these colonies from genuine native stocks.

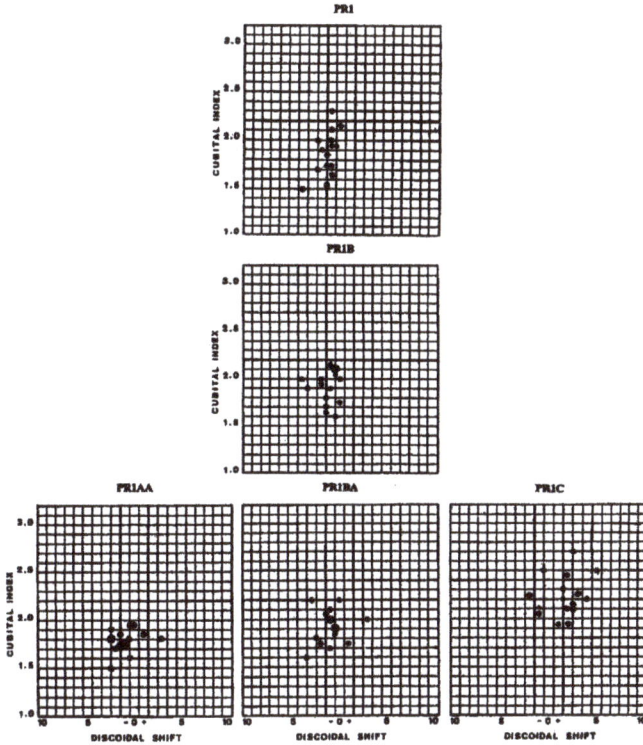

Fig. 12. Back-breeding of the PR strain. Use of wing morphometry to decide between sister colonies PR1AA, PR1BA and PR1C. The line was continued through PR1AA.

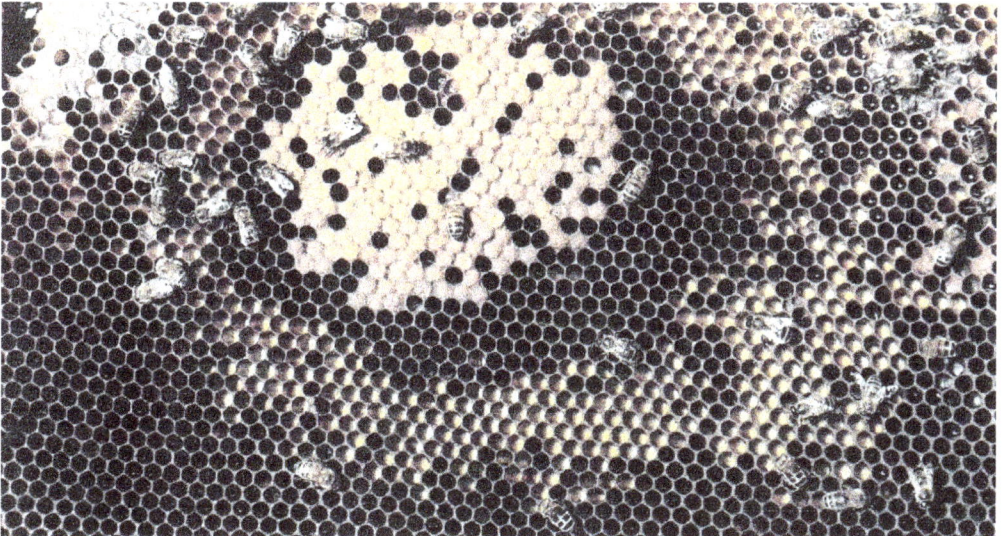

Fig. 13. A brood frame from a native colony showing early stages of pollen storage beneath as well as above and beside the brood.

If you are to conserve the native race, you must not allow any genetic contamination or "genetic introgression" into your reserve area or reserve population. This means you need the cooperation of other beekeepers, which can present difficulties. Nobody likes to be told what to do and beekeepers are especially independently minded people. Winning their cooperation may require tact and generosity on all sides. ***Nevertheless, you must exclude all foreign queens and the drones they produce, or the remaining extant native genome will become hopelessly compromised.***

To achieve true matings of your native virgin queens, you need to propagate a large excess of the best native drones. If a good native queen has mated with foreign or "hybrid" (i.e.cross-bred) drones, she may still be valuable as a drone producer, since drones develop from *un*fertilised eggs. But do not allow badly mated queens to produce laying daughters.

Both virgin queens and drones will fly considerable distances for the purpose of mating within a drone congregation. Kraus (2005) quotes 8 kilometres (5 miles) for queens, and 10 kilometres (6.25 miles) for drones. However, drones will visit several assemblies, and since they need to return to their own hives or another, to stock up with fuel every 20 minutes or so, they probably do not frequently visit distant sites.

When the weather is too cold or wet for flight, native *A. m. m.* queens will mate with local drones at hedge-top level near their own hives. This is called Apiary Vicinity Mating, or AVM, and although this practice enables the queen to perform her main function of continuation of her own line, it has the great disadvantage of increasing inbreeding (see Chapter 4).

In Northumberland virgin queens frequently supersede an ageing mother instead of swarming away to another site. This is probably another adaptation to the frequently poor flying weather and shortage of natural nest sites there.

Siting your hives

Sentimental beekeepers sometimes refer to honey as "bottled sunshine" and it's not a bad description. Without sunshine you don't get, honey and that works both directly and indirectly. Indirectly, sunshine blesses flowers, ushers their progress through the seasons and stimulates their nectar flow. Directly it warms the hive and the scouts and foragers already out on their work, and by being seen through the hive entrance, shows the sleepy occupants it's a lovely day outside and they should get out and get busy.

The sun also does the seemingly impossible in providing a reliable geographical reference point in the ever-changing day, with respect to which the bees can learn and tell others where to go for nectar and pollen. The same sun-oriented behaviour helps guide swarms to their new homes.

Sunshine makes bees happy. They cannot thrive without it, needing five hours of direct sunshine in midsummer or they will not get much honey, and two hours in midwinter. In winter they must work hard to keep warm by vibrating their wing muscles and a couple of hours of direct sunshine from a clear sky gives them the chance to get outside and relieve themselves, then return to the cluster before they get too chilled.

Do not put beehives beside a track used by horses and do not assemble more than twelve in one apiary. Avoid placing your hives under trees, as the rain dripping off leaves can upset the bees. If your hives have plastic roofs, replace those or cover them with metal sheets to prevent occasional falling sticks and branches damaging them.

Place your hives out of the wind and shade, preferably in a suntrap corner and out of sight to passers-by. A couple of the concrete blocks known as "9-inch hollows" make a good solid base for one hive and allows its strapping down against the wind. Place them sufficiently far apart for two additional hives to be placed between during swarm control operations, level them with a spirit level and orientate them so that their entrances face SSE to catch the 10 o'clock sun.

If you place your brood combs "the cold way", that is with their top bars at right angles to the entrance block, the combs are then close to the natural orientation of brood combs, along a SSW - NNE line. I believe the bees construct them in that orientation to keep the brood safe on very hot days, as only the ends of frames are then exposed to the sun's heat at the hottest time of day, around 2:00 pm.

This sounds like an obvious instruction, but I have several times seen beehives standing in bogs that were so wet they sank several inches deep. Never stand hives on wet ground. That is too cold and too humid for bees.

People often ask me where in their garden they should place their new hive. I say: "Is there somewhere where you would sit to have a quiet cup of coffee on a lovely spring morning?" When they say "Yes", I reply: "That's where you should put your hive".

Seasonal climatic change.

Winter temperatures *seem* to present the greatest obstacle for foreign bees in Britain. This is curious as British winters are not particularly severe compared to those in Denmark and Germany, for example (Fig 14a), where some stocks of Mediterranean origin seem to have more success. So why are so many colonies lost here in winter?

Comparison of mean January temperatures with Ruttner's map of the native ranges of the honey bee subspecies also shows no correlation (Ruttner, 1987). But when we compare the latter to July average temperatures, we see something very surprising (Figs. 14b). North of the Alps and coincident with the natural range for *A. m. mellifera*, we see July temperatures are in the range 15-20⁰C, further north, below 15⁰, whereas all the other European races enjoy July home temperatures of over 20⁰ (Fig. 14b).

This tells us that ***it is summer conditions rather than winter that are critical for survival of the Mediterranean subspecies.*** Although the warm-weather bees die off predominantly in winter, the seeds of their demise seem to be sown the previous summer.

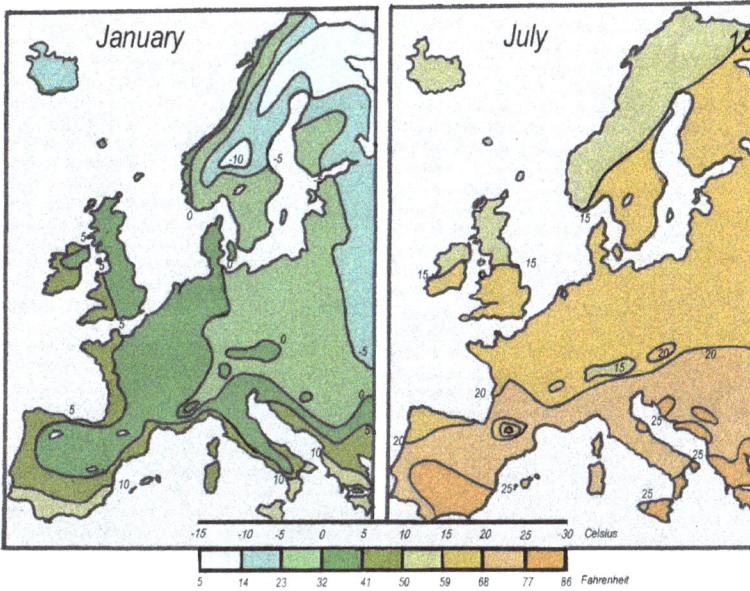

Fig. 14a. Mean air temperatures in Europe as reported in 1973 (Cohen, 1973); a: in January; b: in July.

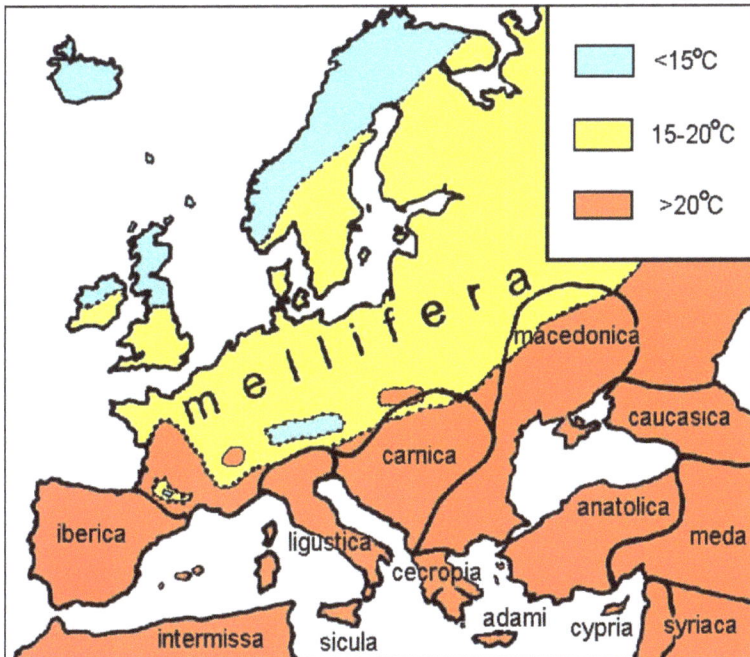

Fig. 14b. Mean July temperatures in 1973 showed a very good match to the natural distribution of the honey bee subspecies as reported by Ruttner (1987). By contrast, January temperatures bear no sensible relationship.

Significant issues associated with cool British summers for Mediterranean bees are mating, foraging and possibly harvesting of nectar and/or pollen from our native wildflowers. Or it could be the designs of those flowers that may inhibit foraging by foreign bees lacking the physical or sensory skills to cope with them.

I am reminded of one of my earliest experiments with bees, when oilseed rape (OSR) first arrived in the North of England. At that time, I had two hives of local dark bees and two of what I now suspect were the original Buckfast (i.e. as Brother Adam designed them, not one of the modern variants). I was new to beekeeping and nervous of losing my hives to theft, so I placed just one hive of each kind side-by-side at the edge of a rape field. OSR had the reputation of producing honey that tasted like turnips, so when the dark colony filled its super, I was overjoyed to find its honey tasted delicious, with no off-tones of turnips! A week later the Buckfast had also filled its super and I expected its honey to be just as good. But NO! – it tasted of turnips!

I initially thought there must be a significant difference in the ways the two strains of bee processed the OSR nectar. However when I looked in the brood nest, I found my local bees had large, yellow pollen stores from the rape, but also other pollen of a variety of colours, whereas the Buckfast had a blaze of yellow and very little else! So my deduction was that, although the different bees might well process nectar in different ways, there was probably sufficient explanation in the pollen colours to explain the difference in honey quality. Evidently the Buckfast bees were less adventurous or less capable of collecting the produce of our wild flowers that the local bees were able to access.

So, it could be that the reason why the Mediterranean derived bees fare less well than natives here is that they are just not sufficiently well adapted to earn a simple living from our flora. Their lack of local adaptation could lead to many deficiencies in essential food components, such as amino acids, vitamins, minerals and possibly essential oils. It is probably an inability to gather sufficient pollen, or a sufficient variety of pollens here that limits their success.

Hazards of winter for Southern bees

Northern winters can kill south-adapted variants of many species, and this applies to the honey bee subspecies that have originated in the warm regions around the Mediterranean, such as the Italian honey bee, *Apis mellifera ligustica*. In their homelands they presumably reduce their activities somewhat towards the later part of the year, but unlike our northern bees, they nevertheless continue to raise brood throughout the winter months in Britain as in Italy.

This is necessary because young brood must be fed on broodfood, produced from the head glands of juvenile house bees. With all honey bee subspecies, as individual worker bees get older they normally lose this capacity and progress to other tasks. But in non-*mellifera* colonies, to secure food for the emerging larvae, there has to be continuous production of young adult bees to ensure continuous production of brood food.

By contrast in *mellifera* colonies the queen reduces her laying rate in late summer and the young workers gorge on pollen and accumulate large amounts of fat in their fat bodies. The physiology of the latter changes along with the need to produce less brood food and one effect of this is that they remain physiologically juvenile. The fat stores are a valuable food resource during winter, so these "winter bees" use less of the pollen and honey stores and have fewer faeces to void outside the hive. Towards the end of winter, the queen begins to lay again, the foragers bring in pollen and the "winter" bees start to produce broodfood again for feeding to the new generation of emerging larvae.

Maintenance of reproductive activity requires considerable heat as well as food for general subsistence and this is why colonies of Mediterranean bees must be kept warm in winter, with top insulation of the hive, and are considered to need feeding very well with sugar in the preceding autumn.

But sugar is not a whole food, and a better policy is to allow the bees to collect their own winter stores of honey and pollen and to leave sufficient on the hive to last them into the following spring. This is preferable because honey collected by the bees is rich in many other necessary food components in addition to white sugar (sucrose). If their natural stores are replaced by pure sucrose they miss out on these other essential nutrients, including those in pollen, and develop an increased vulnerability to disease (see Feeding bees, below).

Enabling the Mediterranean colony to remain active through winter has disadvantages, however. The most obvious is that consumption of stores depletes them, and this may lead to starvation, including "isolation starvation" when the colony moves away from a vital section of the stores and becomes surrounded by empty cells. Here the bees die because workers are unable to move into cold regions away from the brood nest, as when they get cold, they become too torpid to move.

A second consequence of *ligustica* worker bees remaining active and consuming stores is that they need to relieve themselves. For this they must leave the hive when they can rapidly get chilled and die. Outside the hive they can also be eaten by predators such as great tits.

A less obvious disadvantage of maintaining an active brood nest is that it provides a winter habitat for brood diseases such foul brood and varroa. So if the hive has a varroa infestation, the colony has no respite from the mites, which can build up and overwhelm it (see Appendix).

Benefits of winter for Northern bees

By contrast our northern bees have spent tens of thousands of years learning how to cope with cold weather and they do this in several ways, the sum total of which being that they actually benefit from conditions that regularly kill Southerners. This in a way is comparable to the requirement of some northern plants for "vernalisation", or cold exposure, as a necessary precursor to Spring growth.

In late summer the laying rate of the native queen decreases as the income of honey and pollen falls and she eventually ceases to lay (Fig. 28) and if it is given the space the colony will form into a "winter cluster", which behaves in a temperature-responsive way, expanding and contracting to regulate its temperature. The fat worker bees produce few faeces, and these are also abnormal in containing rectal catalase, an enzyme that destroys faecal gas, removing some of the workers' need to relieve themselves outside the hive.

When they begin to feel cold, the workers shiver their powerful wing muscles, without beating their wings. This generates more than enough heat and also maintains their physical fitness, conferring vigour and flying strength when foraging recommences.

Many investigators have confirmed that colonies kept cold in winter have this vigour, noticeably lacking in the same bees kept warm by artificial means such as hive insulation.

The two or three months break in brood production disrupts brood disease continuity, so helping protect the colony against disease build up (Fig. 28). Reproduction of varroa mites is also disrupted, since that requires live brood and if the colony has the ability to kill adult mites by aggressive grooming, the mites cannot hide and fall victim to attack, when most mites are killed off (see "Varroa resistance" and Appendix).

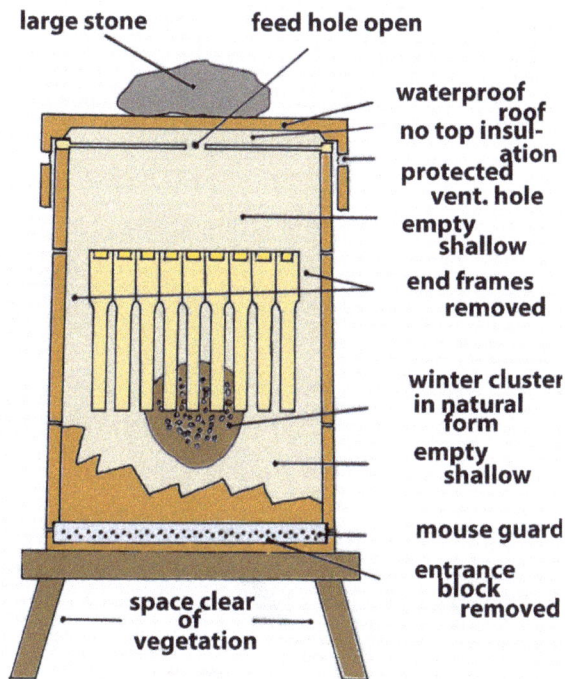

Fig. 16. The winter hive set-up for Northern honey bees.

Seasonal management

Figures 14a and 14b show us that the northern half of Britain has average July temperatures below 15° C and in Chapter 1 I mentioned my frustration at the difficulty of finding a book on beekeeping that is not addressed to those in southern Britain trying to manage subspecies of *Apis mellifera* that by rights should not be there! So there is no difficulty getting advice on how to keep the wrong bees in the South of the country, and you soon find an almost infinite variety of minor differences in how to do it. By "wrong" I mean those of foreign origin and Mediterranean adaptation: July temperatures of above 20ºC.

My experience is limited to northern bees in a region where July temperatures average 14°C, the temperature at which winter clustering is said to begin. I have been taught that at this latitude, the overriding principle is keep your hive occupants warm in summer with added insulation, but cold in winter to force them into their native adaptation.

So in winter we keep our hives cold, without insulation and the door wide open, unless the colony is weak, when we wrap it up warm. The cold set-up also has an empty shallow box below the brood box to allow the colony to hang in a cluster. But in summer we keep both strong and weak colonies as warm as possible. When we see the first pollen going in, in Spring, we abandon the winter set-up and add a sheet of insulating fibre board above the crown board. This is necessary as bees need reliably high temperatures to raise brood and also to secrete wax, in order to build comb and we also need stable broodnest temperatures to avoid chalkbrood. (Cooper, 1986).

The recommended summer set-up is shown in Figure 17 with zero, one or two supers, depending on the abundance of forage, and starter strips for the heather. At other times in the summer, shallow super frames would be fitted with hexagon-embossed wax foundation. Traditionally we have used wired foundation for honey to be extracted by spinning in an extractor, unwired for honey from OSR that is likely to set in the comb and need melting out. The wire is to strengthen combs so they do not rupture in the extractor, but I find this does not happen even with unwired if you use a radial extractor.

The time to add the first super is when honeycomb builds up on the brood frame top bars, and the same for the second super and third. The second is initially added on top of the first then when it is partly filled their positions are reversed, with the heavy one on top and a clearer board inserted between them.

If bees have insufficient winter stores to survive until they are bringing in honey again, then they will usually succumb in March, so be especially vigilant then.

Figure 16 shows the winter set-up for Northern bees, as taught to me by an Edinburgh team. I call it the Scottish method, though I do not know any other Scots who use it. It's important features are that the colony is allowed to regulate its temperature without insulation, but with good air circulation and with space below the broodnest to hang as a cluster. An empty shallow box is placed below the brood nest and another empty above it with no queen excluder between them. The two end brood frames are removed so there is free air circulation around the colony. The entrance block is also removed and the entrance protected with a mouse guard and a crown board with an open feeder hole is placed above. The floor is solid timber, not mesh, and the roof is weighted down with a stone. I have used essentially the same plan for some 20 years, but during that time, have introduced some modifications.

The first is that I do not usually have enough empty shallow boxes, so I usually eliminate the upper one. Sometimes the bees are too fierce at that time of year to remove the end brood frames, so I then leave them in place. Sometimes pygmy shrews squeeze through the mouse guard and kill the colony. If I think that is a risk, I replace the standard mouse guards with shrew guards made by cutting a zinc queen excluder into strips. This however tends to become blocked with dead bees, so needs attention now and again.

If I have to leave a hive unexamined into the spring, the bees may build down from the brood comb bottom bars so, for a happier outcome, I put shallow frames with starter strips of foundation half an inch deep in the lower box. This means I can sensibly move that box into normal use as the spring develops.

I have also replaced the rickety timber hive stand with two 9-inch hollow concrete blocks, which allow me to strap the hive down instead of using the heavy stone.

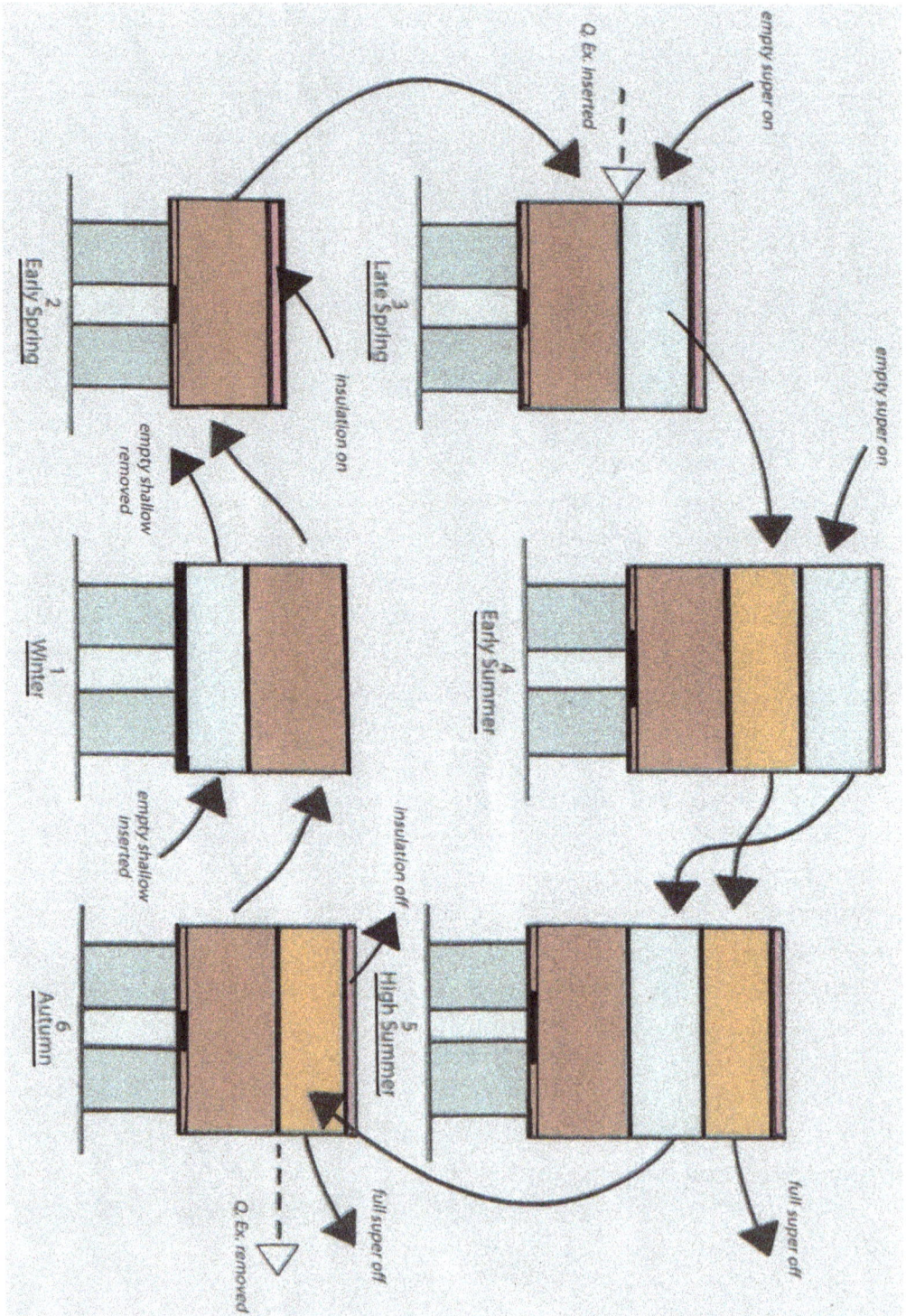

Fig. 17. Seasonal management of the beehive

Figure 17 illustrates the sequence of shallow box manipulations I follow through the year to ensure optimal ease of operation, honey production and colony survival. Swarm management is slotted in during Late Spring and Summer, allowing time for all colonies to build up sufficiently strongly for the Winter.

1. This illustrates the winter set-up as described above, with an occupied deep brood box and an empty shallow box below. The hive entrance is at floor level in this and all other conformations and the size of the entrance is as indicated.

2. When I see the workers taking in pollen usually in February, I move towards summer management by removing the empty lower box and adding insulation between the crownboard and the roof, in the form of a one-inch sheet of expanded foam house insulation, (e.g. Celotex or Kingspan) cut to the size of the crown board and bound at its edges with two-inch tape.

3. I then observe the hive until the brood frames are full, new wax can be seen at the top edges of the brood frames and wild honey comb is being built on their top edges. I then add the queen excluder and super of empty drawn comb.

4. When the first super is full and wild comb is again visible along the tops of the frames I add another super of empty shallow frames or undrawn foundation above the first. A week later, I exchange the positions of the supers and a few days after that, insert a clearer board between them and remove the upper, heavy box a few days after that. Depending on honey flow, I may repeat this operation several times, putting an empty shallow on top, then reversing it with the one below before taking the heavy box away.

5. In late summer if I have the manpower to assist me, I place a super with starter strips of thin foundation directly on top of the brood boxes of the strongest hives and take them to the heather moors. This must be done before August 12th, when the grouse seasons begins. Towards the end of the heather in early September, I clear and remove the supers of heather comb, take the brood boxes backs to the home apiary and ready them for winter.

While at the moors I assess the strength of the broodnest with a view to exchanging frames and uniting for winter.

Feeding bees

In an earlier section I expressed my horror at learning that the recommended artificial food for bees is syrup made from white sugar. This was introduced at least 150 years ago, since when there have been enormous advances in nutritional science. For a long time we have known that all animals require three major components to their diets: protein, carbohydrate and lipid (i.e. fat). In addition, they require minerals, vitamins, common and especially rare amino acids and many other molecules that are present in their normal diets, but certainly not in white sugar, whether from beet or cane.

White sugar, sucrose, is a pure chemical, in fact the purest chemical we, ourselves eat. Somehow few scientists seem to have put their minds to the actual food requirements of honey bees to the extent that their provision has become accepted common practice. So, plain white sugar omits many of the essential nutritional components of honey that bees need to stay healthy.

But it's worse than that. I once worked as a fisheries research laboratory biochemist and in that role I frequently had to look up the dangers of handling unusual biochemicals. It turns out that *everything* is toxic, including water! Toxicologists rarely mention toxic *substances*, but refer instead to toxic *doses* and *everything* has a toxic dose. It is referred to as its LD_{50}. This is the "Lethal Dose" that would kill 50% of a test population of adult laboratory rats if administered all at one time. For sucrose, this is around 30g per Kg body weight, which for a 12 stone (76Kg) man, translates into 2.3Kg. A medium-sized bag of sugar, such as you would buy in a supermarket, weighs 1Kg. So, if a 12 stone beekeeper drank the syrup he could make from two and a quarter such bags at one sitting, he would have a less than 50% chance of survival!

Converting those figures into a suggested toxic dose for a 100mg bee would not be sensible, as there are too many confounding considerations, but I think it makes the point that white sugar is not a normal food that can only do the bees good. The doses of sucrose that are routinely given to bees must be well into the toxic category and could be doing them harm, but I have never seen an analysis that quantifies this possible harm.

Fortunately, however, the suppliers of bee related products, particularly in Rumania, have recently taken steps in the nutritional direction, and it is now possible to buy imaginatively formulated syrups and fondants, that for example "represent the complete solution between nectar and pollen", being fortified with plant extracts, vitamins, amino acids, proteins and other sugars. There are claims to promote intestinal well-being with probiotics, including lactic acid bacteria that actively fight disease.

Such formulations could be invaluable if nectar and pollen dearths cause significant problems for the bees.

Global warming

With current concern about global warming, a pertinent question is: *"How will global warming affect the native honey bee?"* I am told that the expected rise in global temperatures, if present meteorological conditions persist, would be in the range of 2 - 3°C.

According to the July temperature map (Fig, 14b), July temperatures near the city of York at Latitude 54° North, in Northern England, average 15°C. Literature produced by the Scottish Beekeepers' Association indicates that the northern limit for wild or feral colonies in Britain is just north of Inverness and there July temperatures average 13°C. Inverness is around 300 miles north of York, so a 2° Celsius drop in average July temperature corresponds to something like 300 miles in the South-North direction.

If global temperatures rose by 2 or 3 degrees, average July temperatures in York might be expected to go up to 17 or 18°C and those in Inverness to 15 or 16°C and the northern limit for feral colonies might then move up to Orkney.

In the South, a 2° or 3°C rise might allow *A. m. iberica* to move north from Spain into France and possibly even Cornwall and the Scilly isles, but *A. m. ligustica* and the other Mediterranean subspecies would still probably not be able to survive unaided and reproduce in Britain.

A warmer climate should allow native bees to survive at higher altitudes, but movement of bees would only follow colonization of uncultivated land by low altitude wildflowers and of cultivated land by agricultural crops like sunflowers, that require relatively warm conditions.

However, things are much more complicated than that and the above comments should be considered as being relevant to the near future only. This is because temperatures in Britain are influenced also by another major heat source, the Gulf Stream. This is a warm current that picks up heat in the Gulf of Mexico and delivers it to our western shores.

The Gulf Stream is a part of what geographers call the Atlantic Meridional Overturning Circulation (AMOC) or "the North Atlantic Conveyor". The basic issue here is that movement of warm surface water out of the Gulf of Mexico is powered by the sinking of particularly dense surface waters in the Arctic made dense by cooling and the freezing of sea ice, a process that expels salt and makes the surface water unusually briny. Being cold and heavy, it sinks to the sea floor and is replaced by lighter, warmer water from the tropics. The warm water the AMOC carries north releases heat to the atmosphere and plays a crucial role in keeping Western Europe warm. It has been estimated that winters in the UK (lying at the same latitude as Alberta, Canada) would be around 5° C cooler if it made no contribution (Ref.: Climate impact gulf stream, 2024).

This power source wields about 50 times the energy use of all humankind, but is now under threat, due to the loss of Arctic sea ice. This cuts off dense brine production, supplemented by a substantial increase in light, fresh water running off the Greenland ice sheet as it retreats, both consequences of climate change. The power of the AMOC is now calculated to be 15% weaker than it was in 1950, or at any other time in the last 1600 years.

The sensitivity of the AMOC to climate change through historical times is estimable from the size of sediment grains in samples drilled from the ocean floor: the larger the grains, the stronger the current at the time of deposition and *vice versa*. A reduction in the force of the current is evident from about 1850AD, coinciding with the end of the Little Ice Age and increasing precipitation over the subpolar ocean and neighbouring lands. In a process that foreshadowed what we are seeing today, this increased the freshwater input to the northern Atlantic, melting massive quantities of ice that had built up during the Little Ice Age, reducing the density of Arctic surface waters and their tendency to sink, and so ultimately weakening the northward flow of warm water from the Gulf.

A key feature of the AMOC is that it relies on a feedback loop: warm salty water from the tropics becomes cold salty water in the Arctic, which sinks and brings in more warm, salty water. While the AMOC flow is already predicted to slow down by up to half, this century, if the feedback loop breaks, the AMOC *could* switch off entirely. There is a lot of scientific uncertainty around this climatic 'tipping point', but if passed, it would have unpredictable consequences that could be devastating (Ref.: Is the AMOC approaching a tipping point? 2024). It's cooling effect may counterbalance the UK summer warming described above, but it would also greatly increase the temperature gradients across Europe, leading to "unprecedented" storms, significant drying and much colder winters.

Glaciologists warn us that 'tipping' of the AMOC system has happened in the past and led to "some of the most dramatic and abrupt climate shifts known" (Ref.: Is the AMOC approaching a tipping point? 2024) and could lead to landscapes looking quite different from today.

Our native honey bee survived the Little Ice Age and may be able again to adapt to both possible increases and decreases in temperature that could result from "global warming". But climate change is not just a change in temperature: it also involves rainfall, humidity and wind, etc. and extremes in all of these would seriously disrupt the ecosystems of all pollinators. We cannot predict consequences if the AMOC were to be pushed past its tipping point, but it would certainly be very much the best thing if human behaviour never causes or allows that to happen!

Chapter 4

POPULATION CONSIDERATIONS

Worldwide threats to honey bee survival

On a World scale, honey bees are in a very bad way! This applies to many pollinator species, but especially so to Western honey bees, i.e. the several subspecies of *Apis mellifera*.

The COLOSS survival survey (Büchler *et al*, 2014), was set up to compare survival rates of imported versus local honey bees in 21 professionally run apiaries across mainland Europe when all medication was withdrawn. Management practices were standardised, and the experiment was continued for 900 days.

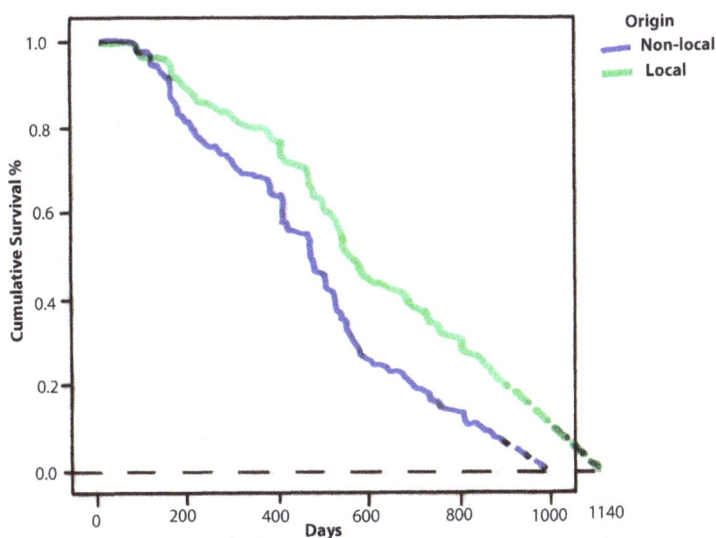

Fig. 18. The COLOSS Survival Survey (redrawn from Büchler *et al*, 2014). The blue, lower line shows results for non-local species, the green, upper line those for local, locally adapted ones.

Fig. 18 summarises the COLOSS observations, the blue line representing the foreign bees, the green line those of local origin. It reveals that local bees can indeed survive for significantly longer periods than imported stocks, as most of us would expect and indeed at every one of the 21 sites. If the graphs are projected to cut the "zero" survival line, i.e. the times by which all colonies probably would have died if the experiment had not been terminated, that would have been at roughly 1000 days for the imported and 1140 (3 years 45 days), for the locally adapted.

If these colonies are a representative sample (which we assume they should have been) the data seem to indicate that only a very small percentage of mainland European colonies can survive for even **THREE YEARS** without medication!

When you consider that practically all locally adapted colonies would survive indefinitely in Northumberland, for example, only 30 years ago, this is a shocking observation and one that terrifies me! Ian Campbell, the then News Editor of "*BBKA News*" reported in 2022 that a more recent North American study in five US states revealed the over-winter survival rate there for commercial stocks without medication to be a shocking 3%! Of course, *Apis mellifera* is not native to the Americas, but for comparison, my own native and near-native bees have not received medication since 2002 (i.e. for 22 years) and their survival rate for both the last two winters was 90%!

The US National Honey Board reported that in 2008 the US had 2.67 million colonies, while Meixner *et al* (2010) calculated that the vast majority of these were descended from just 500 queens! That's more than 5000 descendent colonies per queen!! There is surely a very important message here! (See: "The most adverse consequences of breeder queens", below.)

Genetic variation and the extinction vortex

As Nils Drivdal has pointed out (N.D., personal communication), all genetically uncompromised *native* species have the secret survival weapon of *genetic variation* and it is more than just possible that the poor survival of commercial stocks in both the US and Europe is because beekeepers and bee breeders, operating on an international scale, have destroyed or lost that vital secret weapon!

Genetic diversity is undermined particularly by three things:

1. Reduction in size of the population,

2. Inbreeding,

3. And in the case of honey bees, multiple queen production using "breeder queens", with the vast majority of the parental generation of queens not breeding.

What is more, these three tend to follow in the sequence shown. First is reduction in population size, which could have many causes, but raises the problem of both queens and drones finding unrelated mates, hence inbreeding. Then further problems consequent upon inbreeding come into play, such as recessive disease, and increased production of diploid drones (see below). The latter leads to reduced brood viability and overwintering failure. Beekeepers then respond by propagating more queens and the favoured way of doing that is by identification of their few best queens and their use as "breeders", to produce multiple daughter queens. But this strategy further exacerbates the original problem (see below).

Apart from the issue of breeder queens, this same pattern has been shown to have occurred with many now extinct species as they each approached extinction. It has been described as an "extinction vortex", as once entered, the situation gets progressively worse and the species is unable to escape. I visualise it as sitting in a kayak skirting a powerful whirlpool. Initially, if you recognise the danger you may be able to paddle to safety. But if you don't, you get sucked ever deeper into the swirling water and will probably never escape.

The term "extinction vortex" was first coined by Gilpin and Soulé (1986) though the concept was also voiced by Caughley (1994). Fagan and Holmes (2006) investigated 10 time-series of vertebrate species that actually went extinct: two mammals, five birds, two turtles and one fish, and from them were able to discern and mathematically define common patterns. Brook, Sodhi and Bradshaw (2008) discerned synergies and stated: *"owing to interacting and self-reinforcing processes, estimates of extinction-risk for most species are more severe than previously recognised"*. They advised that *"future work should concentrate on how climate change will interact with and accelerate ongoing threats to biodiversity, such as habitat degradation, over-exploitation and invasive species."*

Adoption of the "breeder queen" approach to honey bee queen propagation could be the ultimate fatal step in their progression. Below I explain how that could theoretically reduce their over-winter survival to the point of extinction.

The ultimate limiting factor, *csd* polymorphism

In 2005 F.B.Kraus made the astonishing claim that *the ultimate limiting factor determining the viability of an isolated population of honey bees is the number of its sex determining alleles!* To me that statement was astonishing, as at first there seems to be no link between sex determination and colony viability. But if such a connection can be established, we can perhaps refine our quest for the responsible aspect of genetic variation, to loss of polymorphism of the gene that determines sex!

Before we try to explain the thinking behind Kraus's statement, let us consider what we need to know about the genetic determination of sex in honey bees.

The *Complementary Sex Determiner* is the gene responsible for the differences between male and female development in bees and some other insects, given the abbreviation *csd*. Normal males have only one set of chromosomes (i.e. are *"haploid"*) and so have only one *csd* allele, *a1, a2, a3*, etc. and this defines their masculinity.

Females have two sets of chromosomes (i.e. are *"diploid"*) and always have two different *csd* alleles, e.g. *a1/a2, a3/a7*, etc. All female bees are therefore *csd heterozygotes* and that defines their femininity.

Csd homozygotes, e.g. *a1/a1, a5/a5*, etc. are also diploid, but begin their development as abnormal males, known as *"diploid drones"*. In the natural situation these are always destroyed by the nurse bees soon after they emerge from the egg, leaving empty, uncapped brood cells among the worker brood. In some cases this deficiency of workers can reduce "brood viability" to the extent that the colony collapses.

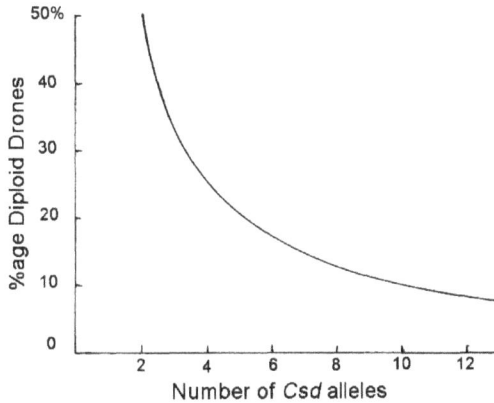

Fig. 19. Relationship between percentage of diploid drones in an outbred population and the number of *csd* alleles.

As shown in Fig. 19, the percentage of diploid drones is very high when the polymorphism of *csd* is very low, i.e. when there are few different *csd* alleles in the population. This curve is described by the equation:

$$\% \text{ brood viability} = (n\text{-}1) \times 100/n$$

where n is the number of *csd* alleles (Mooney, 2018).

Fig. 20 (Woyke, 1976) shows the same graph inverted to emphasise how increasing polymorphism of *csd* is positively related to worker brood viability. This is the first indication we have that a large number of alleles of the sex determining gene (i.e. its high polymorphism), carried by drones in a drone assembly has a favourable influence on the health of colonies led by the queens that mated there.

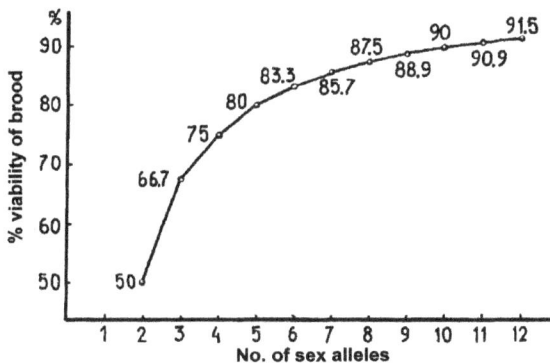

Fig.20. Relationship of worker brood viability to number of *csd* alleles (Woyke, 1976).

This surprising conclusion leads to recognition that ***polymorphism of the csd gene must be considered one of our highest priorities in honey bee conservation. Everything must be done to retain that particular genetic polymorphism.***

We will now take this idea further, to quantify the effect and investigate why and how, despite its widespread use, the practice of queen propagation from "breeder queens" may not be at all a good idea!

Loss of fertility and fertility genetic variation through use of "breeder queens"

Virgin queens have been reported to fly up to 8 kilometres to a drone congregation area and drones up to 10 (Kraus, 2005). But imagine you have 10 colonies in an isolated apiary outside those ranges. You feel that your colonies are not outstanding except for one that is markedly superior and you are persuaded that if, instead of allowing them to reproduce when and as they see fit, you should produce a number of daughter queens from that best one and distribute those among the other colonies. This strategy, you are convinced should greatly improve the quality of your apiary as a whole.

You recognise that is how other livestock breeders operate with cattle, sheep, poultry, and goats, etc. the only obvious difference being that those animals are typically bred from single, outstanding males. So it looks like a "no-brainer", the clever thing to do.

You have inherited a Jenter cage that should allow you to produce multitudes of daughter queens and you decide to give it a try. So you clean up the cage, place your best queen inside, wait till she has laid it up, then release her. You take 15 plugs from the brood comb wall of the cage, each holding a newly emerged larva, mount them in appropriate plug holders and suspend them vertically in a modified brood frame. You then place the loaded frame in a newly de-queened colony where most of the larvae then develop into young queens.

Let's say that 10 queen cells are well drawn out and when sealed you place them individually in queenless mating nucs. You leave these around the apiary to emerge and get mated by the drones from your other hives.

When the new queens have mated and are laying in the nucs, you remove the old queens from your 9 old hives and replace each with a new queen of the "F1 generation". Everything goes well that season. Their brood nests have a few empty worker brood cells, as is normal for a variety of reasons.

The next summer, these F1 queens themselves reproduce, raising queens that mate with the drones produced by their sisters to yield an F2 generation. But this is when things start to go badly wrong, as you see large numbers of empty brood cells. Production of worker bees is reduced, as is honey production and several colonies do not survive the winter.

What has gone wrong?

An explanation

Such an outcome has a straightforward explanation in the context of the *csd* gene and diploid drone production. (See Appendix 1 and Chapter 6.)

If there were 10 or 11 original queens that had come from different apiaries, between them they could have had 20 different *csd* alleles and their drones collectively would be producing sperm of the same 20 different *csd* types. The breeder queen herself has two *csd* alleles, let's call them *a1* and *a2*, and would have produced just two different types of ova, genotypically *a1* and *a2*.

If there were 10 F1 daughters mating at random in the apiary, there could have been a total of 12 or so different types of sperm and ova, but half the daughters would have the *a1* allele from the breeder mother and half the *a2*.

The probability of a daughter being of genotype *a1/-* would be around 50% and *a2/-* also 50%, the other allele in each daughter being one of the other 10 or so types of alleles carried by the drones with whom their mother mated.

The ova produced by that F1 generation of queens would also be expected to be approximately *25% a1*, 25% *a2* and 50% one or the other of the *csd* alleles *a3- a12*. The same proportions of the different *csd* alleless would be present in the drones those queens also produced and in the sperm produced by them.

Assuming there is no striking difference in the vigour of the drones the relative probabilities of the different zygotes in the following F2 generation would be something like those shown in Appendix 1 which illustrates a total of 400 outcomes. Of these, about 25 would be expected to be *a1/a1* homozygotes, 25 *a2/a2* homozygotes and 10 homozygotes of the 10 other *csd* genotypes. The total probability of homozygotes would therefore be ~(25 + 25 + 10 =) 60 out of the 400, i.e. ~15%. That of heterozygotes would be ~ 340, i.e.~ 85% of the total.

The homozygotes would all develop as diploid drones and be eliminated, causing an estimated overall decline in fertility of the apiary of about 15%, so the brood viability would fall to around 85% of that of a comparable outbred colony.

A similar loss of worker bees would be expected to continue into subsequent generations and the outcome is likely to be losses of overwintering colonies, as explained in Chapter 4.

Those ova that were fertilised, but not by the sperm of a brother would include *a1/a2, a1/a3* .... *a1/a12* and *a2/a1, a2/a3* ...... *a2/a12*, etcetera and there would be a persistent preponderance of alleles a1 and a2 in the apiary.

After the F2 generation, if no additional genetic material were introduced, I would expect the outcome to be comparable, with a persistent loss of a large proportion of the worker brood as diploid drones.

Brood viability and over-winter survival

In a complex trial that involved setting up colonies with different numbers of *csd* alleles, Tarpy and Page (2002) found that survival of their colonies over winter required worker brood viabilities of at least 75%, i.e. with fewer than 25 empty worker brood cells in every 100.

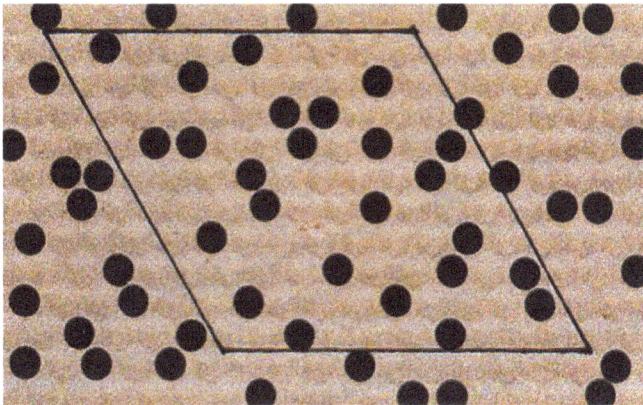

Fig.21. This is what 75% brood viability looks like. The rhombus delineates 100 cells, of which 25 are dark, as the larvae they contain have died.

Fig. 21 illustrates what 75% brood viability looks like: within the rhombus template circumscribing 100 worker cells, 25 of those cells are empty. ***For over-winter survival, you need to ensure the density of holes in the brood is no greater than this***.

Some of these gaps would probably be due to diploid drones caused by *csd* homozygosity. Others could be for such reasons as age of the queen, egg infertility or larval failure, as these increase with maternal age.

We have already seen Woyke's graph (Fig. 20) showing the percent viability of worker brood plotted against the number of *csd* alleles in the population. Fig. 22 shows an application of this representation. This also includes Tarpy and Page's 75% borderline for over-winter survival, as well as the theoretical graph for brood viability for 1-2 year old queens adjusted by addition of 9% brood loss. The latter adjustment is indicated by experiments on brood viability in relation to queen ages (Al Lawati and Bienefeld, 2009). The two lines cross at a point corresponding to just above 6 sex alleles.

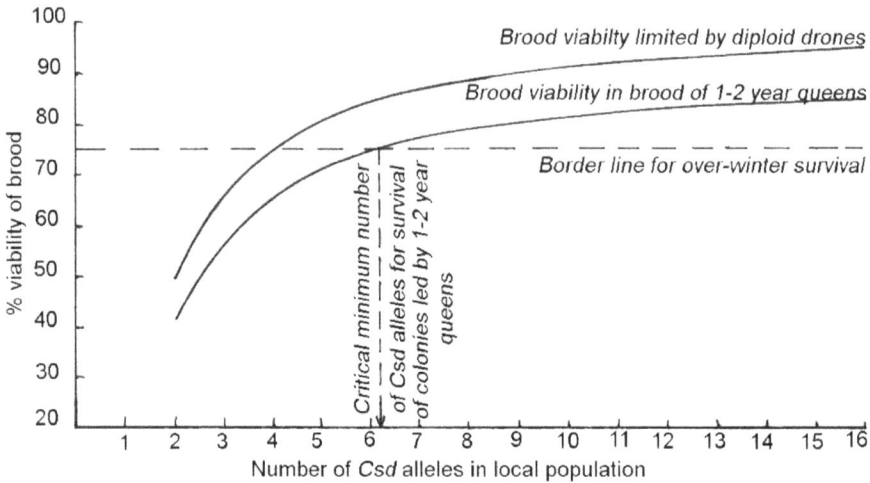

Fig. 22. Prediction of colony survival prospects from brood viability. This diagram represents the brood viability limited by diploid drone formation only plotted against the number of Csd alleles in the local population and the prediction for a queen in her second year with the appropriate additional age-related reduction in fertility of about 9%.

What this means is that if there are fewer than 6 sex alleles in the population, there would be sufficient diploid drones formed that in combination with the brood infertility of 1-2 year queens, would put the colony into the danger zone for winter survival. Colonies then die because the broodnest lacks the capacity to generate sufficient heat for survival. A younger queen or a queen mated by drones carrying one more *csd* alleles would push the picture slightly to the right, where the over-winter survival rate was reported by Tarpy and Page (2002) to be 100%!

Fig. 23 shows this distinction more dramatically: fewer than 6 *csd* alleles goes with a below 40% survival rate; more than 6, with 100% survival! There is apparently a tipping point at around 6 *csd* alleles (and 25% diploid drones), above which survival is practically guaranteed, but below which survival is uncertain and averages out at just below 40% (actually 37.2%, Tarpy and Page, 2002).

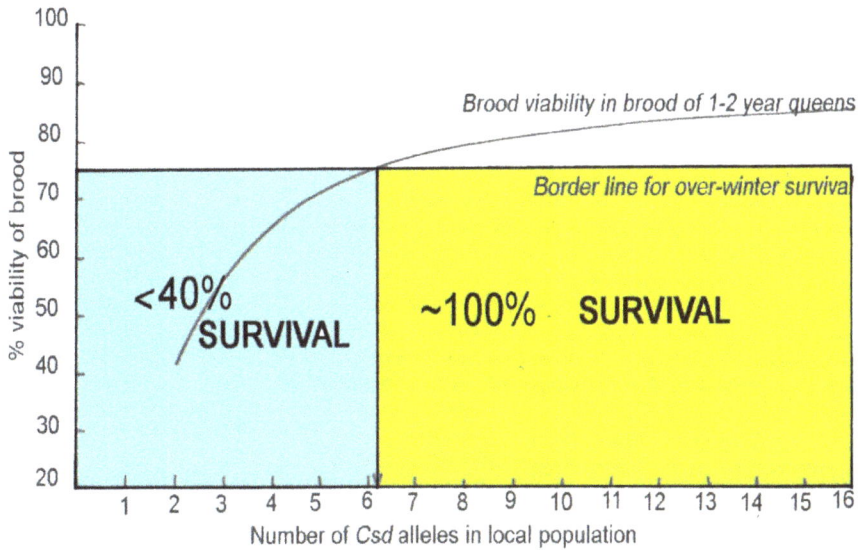

Fig. 23. Winter survival and number of *csd* alleles. This diagram represents brood viability for a queen in her second year, incorporating both the effect of diploid drone creation and reduction in fertility of about 9%. Due to the age-related effects. (Al Lawati and Bienefeld, 2009).

Those figures, derived by Tarpy and Page, must vary depending on the strain of bees, hive type, preparatory feeding, the prevailing climate and the weather at the time, but the experiment should be sufficiently robust for similar general conclusions to be drawn.(See also Chapter 6).

Strength of the broodnest

The strength of the broodnest depends not only on the viability of the brood, but also on the number of frames containing it. My practice has always been to assess broodnest size when the bees are at the heather, in August. The queens are usually in lay then, sometimes again, having had a brood break in July.

I have found that colonies in timber National hives with solid floors and 8 or more frames of brood in August have no difficulty surviving a Northumberland winter. Those with 6 benefit from supplementation to bring it up to 8, while those with only 5 are at risk and best treated as nucs, with extra insulation on the outside of an expanded polystyrene nucleus box. Colonies smaller than 5 frames of brood should be united, preferably with larger ones.

My bees are near-native or native and the precise values I have found would not be expected to apply to bees of other races.

It would be wise policy to ensure all one's colonies going into winter meet all three requirements: 8 occupied brood frames, >75% brood viability and at least 6 csd alleles in the population.

The most adverse consequences of breeder queens

Use of breeder queens is popular among bee breeders as impressive results can be produced easily and rapidly, but as shown above, a longer view reveals serious weaknesses. The example given is on a very small scale and assumes the mating apiary is isolated from other hives. In common commercial practice thousands or even tens of thousands of queens may be produced from one outstanding queen. A beekeeper buying several queens from a commercial supplier may have therefore unknowingly obtained closely related individuals, though to guard against inbreeding hazards, one hopes that mating would have been done free, with a profusion of unrelated drone-producing colonies.

The long-term outcome of such commercial practices depends on the locations where the newly mated queens are distributed, whether that apiary received queens from the same supplier previously and the health status of bees in the vicinity of the recipient apiary.

But we have not yet considered the most negative outcome of a British beekeeper using a breeder queen selected from perhaps 10 in his own apiary, as described above. This concerns the daughter-less status of the 9 other queens. *If they are not bred from, their genomes are effectively thrown away and the valuable genetic variation they had is largely lost! In fact, 9 queens out of 10 not being used for breeding would constitute a 90% loss of that potentially very valuable genetic resource they carry.* True, their drones carry the same information, but once those queens have been replaced, their drones and the multitude of genetic variants they might have been carrying, drop out of the picture, or their frequency is greatly reduced. In the present state of honey bees worldwide, *such wholesale loss of genetic information is definitely something we cannot afford!*

From a conservation viewpoint, therefore, *I believe breeder queens should not be used* unless corresponding steps are taken to preserve *all* the potentially valuable genetic variation carried by their sisters.

For conservation of local genomes I suggest the best policy is to allow all native colonies to reproduce without human-guided selection, allowing nature to make her own choice of survivors.

Can a bad situation be saved?

The answer is probably yes! If the number of sex alleles in a population looks hopelessly low, it may nevertheless be capable of improvement and the population saved. Based on his experience with the closed population on Kangaroo Island, Woyke (1976) recommended: *"The best way to increase the number of sex alleles in a population and to increase the survival of brood, with a minimal change of other characters, is to introduce one or a very few queens inseminated by several (not many) different drones."*

Whether or not the lost background genome could also be restored would then depend very much on the choice of drones used.

Summary: to ensure over-winter survival of your colonies:

▶ Replace queens older than 2 years.

▶ Keep hive populations and regional populations large.

▶ Raise queens and drones from several queen lines.

▶ For mating, transfer your virgin queens to another apiary populated by drones of other queen lines.

▶ Ensure there are 7 or more good frames of worker brood per colony in August; supplement weak colonies with sealed brood from other hives.

▶ Check brood viabilities with the 100-cell template when setting up for winter.

▶ Set up your hives with an empty box below the brood to allow the bees to cluster (see Chapter 3).

▶ Remove the entrance block, but screen the entrance with a mouse- or shrew-guard.

▶ Remove top insulation, so that bees are forced to adopt the winter cluster conformation (Chapter 3).

Chapter 5

INHERITED ATTRIBUTES AND BEHAVIOURS OF NATIVE BEES

Brother Adam and the Buckfast Bee

One of the greatest and certainly the most famous British beekeeper was Brother Adam of Buckfast Abbey in Devon. He was born Karl Kehrle in a German family and at the age of 12 was conscripted by the Abbey with a view to training as a stonemason, to help build their splendid new abbey building. The construction gang was small and brother Adam grew into a small youth, lacking the physical strength to manoeuvre heavy blocks of stone. So, he was instead transferred to the apiary, to assist with practical beekeeping.

In that he became a master and remained a devoted beekeeper to the end of his days. But his initiation was difficult, as less than a year later, in 1915, the apiary was hit by the honey bee epidemic known as Isle of Wight Disease (IOWD).

The cause of IOWD was not completely explained for half a century, but it now seems to have been caused by Chronic Bee Paralysis Virus (CBPV) vectored by acarine mites that infest the tracheae (breathing tubes) of honey bees (see below). This devastated apiaries countrywide and exterminated all races of bee in the Buckfast apiary except those of the Ligurian variant of the Italian bee, *A. m. ligustica*, and its first-generation hybrids with the native British bee, *A. m. mellifera*. A factor in that resistance was considered to relate to the growth of bristles obstructing the entry of acarine mites through the spiracles at the outer

61

ends of the tracheae. This selective resistance gave Brother Adam the motivation to establish an IOWD resistant strain, that became known as the Buckfast Bee.

In its development over the subsequent 70 or 80 years, Br Adam visited many remote areas of Europe, brought queens back with him and tested their properties with a view to building up a superlative strain of bees with the best traits of many varieties, in addition to resistance to IOWD. This much extended operation is described in "*In Search of the Best Strains of Bees*" (Adam, 1983) and "*Breeding the Honeybee*" (Adam, 1987).

A vital aspect of the operation was the comparative evaluation of eventually 10 bee stocks for their relative performances, that also incorporated assessments made from as long ago as 1898. These dealt with 16 traits: industry, resistance to brood and adult diseases, disinclination to swarm, longevity, wing power, hardiness, keen sense of smell, readiness to enter supers, comb building ability, good temper, calm behaviour, propolisation, absence of brace comb and sense of orientation. The comparative performances of each of the stocks are recorded in Adam, 1987, as a table of scores from -6 to +6.

To the modern reader, there seems to be a notable omission from that table in that there is no mention of *A. m. mellifera* by that name and this has, I believe, given rise to the erroneous belief that Brother Adam did not think highly of our native, dark northern honey bee. That apparent omission is however due to a temporary change in the Latin name of the North European native bee!

Our northern bee was originally named by the Swedish naturalist Carl von Linné (Linnaeus), the word "*mellifera*", being Latin for "*carrier of honey*". However, some biologists considered it inappropriate, as bees usually do not carry honey, but the nectar they take from flowers and later convert into honey. The name "*mellifica*" was therefore substituted for "*mellifera*", meaning "*maker of honey*" and that is the name in Br Adam's table (Adam, 1987). Subsequently this nomenclature change was reversed in accordance with a new rule that the first ascribed name should be retained, but Adam's table was not altered accordingly.

An additional complication arises from a one-time suspicion that the honey bees of Northern Europe may be of two varieties. One of these is the one that lives in woodland, given the varietal name *sylvarum* by Goetze (1964), the other on heathland and called *lehzeni* by Büttel-Reepen (1906; see Wikipedia: "*Western honey bee*"). The name used by Adam in his table was "*Mellifica lehzeni* meaning "*heathland honey maker*", which despite its name was actually the ubiquitous *A. m. mellifera*. Careful examination of the two ecotypes revealed no consistent differences and that distinction has now also been abandoned.

Br Adam's scores for "*Mellifica lehzeni*" are +6 (i.e. the highest) for seven of the 16 listed properties, but -5 for both "good temper" and "calm behaviour", and -3 and -1 for resistance to brood and adult diseases respectively (see below).

More recently, Hubert Guerriat (2013) produced a comparison of the four most common honey bee subspecies in Western Europe: *A. m. mellifera, carnica, caucasica* and *ligustica* as well as Buckfast. This was based on 10 studies performed independently by bee scientists distributed widely across the native range of *A. m. mellifera* (Fig.14b) and included two comparative evaluations by Brother Adam. An adaptation of the same data is shown here in five behavioural categories.

I have also quantified Guerriat's "traffic light" classification by representing his green, amber and red categories by numbers. The first of these comparisons is shown below in relation to flight and foraging.

Gentleness

Brother Adam's (1987) rating of "good temper" of *A. m. mellifera* as "-5" demands a response and I note also that one often hears or reads the opinion that "dark" bees (usually equated with *A. m. mellifera*) are highly aggressive. This is not at all the case and indeed Beowulf Cooper, the greatest authority on the native bees of Britain (Cooper, 1986) wrote that aggression in *Apis mellifera* is the best indicator of inter-subspecies hybridity!

There are two issues here, both related to genetics. The first is the observation that the honey bee races are classifiable with regard to inter-race compatibility of temperament (Cooper,1986). The M-lineage, of which the major member is *A. m. mellifera*, constitutes one temperament group, bees of the C-lineage, of which the major members are *A. m. carnica* and *A. m. ligustica*, are in another. Crosses within either group do not usually affect temperament in the offspring, but crosses between members of different temperament groups can have unpleasant outcomes, although these usually do not become apparent in the first or F1, generation, in which "hybrid vigour" is also a factor. In these crosses aggression arising in later generations is common and can be severe. This is sometimes called "F2 aggression", implying that it occurs in the "grand-offspring", although it can appear several generations later. A cross between dark *A. m. mellifera* and dark *A. m. carnica* bees could therefore yield highly aggressive descendants and be indistinguishable from pure *A. m. mellifera* at a casual glance. The outcome of crosses between *A. m. mellifera* and Buckfast are notorious for their aggression.

The second issue relates to body colour. A Northumberland farmer and personal friend, named Colin Weightman, crossed yellow Italian bees with local dark bees, with the aim of taking advantage of hybrid vigour in the F1 generation. According to Mendelian genetic reasoning, in such a cross you might expect the offspring to be all of similar appearance, i.e. all dark, or all yellow, or all of some intermediate shade, but he found something unexpected. When the cross was performed between dark queens and yellow drones, the hybrid offspring were all dark, but when the queens were yellow and the drones dark, the offspring were all yellow! Again, on Mendelian reasoning, there should have been no difference between the outcomes of the two mating systems, but here is an example of an "epigenetic" effect overruling the expected predominant influence of the underlying pigment genes inherited from their parents (Weightman, 1961).

By exchanging crossbred eggs between yellow and dark colonies, it was deduced that it was the physiology of the nurse bees in the rearing colonies, that was mainly responsible for defining the colour of the developing worker bees. This was explained as being due to a "nursing factor" in the brood food defining the young "hybrid" bees' pigmentation, rather than the pigmentation alleles they had inherited from their parents (Rachel Lowther, personal communication).

Weightman's experiments showed that body colour is not a reliable indicator of worker bees' genotypes or genetic background and it is for this reason that when trying to define honeybee subspecies, we need to resort to other methods, such as wing morphometry or DNA analysis.

Reports relating body colour to aggression should therefore not be taken seriously unless the bees concerned have been identified by better indicators than body colour. In my experience, those who manage unequivocally pure-bred native *A. m. mellifera* generally vouch for their exceptional gentleness.

Breeding for gentleness

Bees with a strong tendency to sting take a lot of the joy out of beekeeping. They also discourage close inspection, so that disease and swarm preparations are more likely to pass unnoticed. So, in my opinion even in the context of conservation, honey bees should be rigorously selected for non-stingy behaviour.

Brother Adam claimed that it is easy to spread gentle behaviour to other hives by propagating extra drones in gentle colonies, which will then mate with many virgin queens and cause their daughters to be less defensive (Adam, personal communication). I have found the same. My back-breeding exercise (Chapter 3) produced bees with apparently all the characters of natives, except that they were rather sharper than I liked. However, one swarm I acquired from Cragside in north Northumberland was exceptionally gentle. This in fact was the first real native stock I found, as indicated by its mitochondrial DNA, which was identical to that of Andrew Abraham's bees on Colonsay (Bo Vest Pedersen, personal communication). Propagating its drones converted all my next generation of bees into benign ones. A neighbouring beekeeper was mystified about the transformation of his long-term nasty bees into nice friendly ones. Mine became so gentle that I ran several apiary meetings without lighting my smoker while working throughout with bare hands.

Sometimes when you leave the vicinity of the hive you have been inspecting, guard bees will follow you to a considerable distance. This is called "following" and may lead to adverse criticism from neighbours and members of the public, as well as visitors to your apiary, so "following" also should be selected out.

Some bees have a tendency to run across the comb, or to jump. These take longer to inspect, when chilling can reduce brood survival. Selection for "sedate" behaviour makes for easier handling.

Foraging and flight

Fig.24 shows an adaptation of Guerriat's (2013) report on honey bee behaviour in which just those scores relating to flight and foraging are presented. The vertical columns represent from left to right *A. m. mellifera, carnica, caucasica, ligustica* and Buckfast and their performance in each of six aspects of flight is shown as green for the best, amber for an intermediate level of performance and red for the poorest.

The overall scores for each genotype are shown below: green being awarded 3 points, amber, 2 points and red 1. It is revealed that for flight and foraging activity they earned **18, 11, 6, 5** and **6** points respectively. But that score overlooks the fact that two of the subspecies were assessed for all six attributes, one subspecies for five and two for only two attributes. A correction was therefore applied by dividing the first scores by the number of appropriate assessments, to give the normalised scores shown in the lowest line of figures: **3.00, 2.20, 1.20, 1.25** and **1.05**. That is a better estimate of their relative strengths overall and shows that *mellifera* is by far the strongest, *ligustica* by far the weakest and the other three of intermediate strengths.

The qualities assessed cover motivation, as "keenness to forage", sensory skills as "sense of orientation" and "sense of smell" and the physical aspects of foraging as "flying strength", "flight activity" and "flight activity at low temperatures". Those attributes are also reflected in other recognised race specific characteristics, such as wing shape, as described by the Cubital Index and Discoidal Shift (see Chapter 2) and "Acclimatisation" (see below).

INHERITED ATTRIBUTES AND BEHAVIOURS, Table

	M E L L.	C A R N.	C A U C.	L I G U.	B U C K.	References
Foraging and flight						
Keenness to forage	O	O	O	O	O	Br. Adam, 1985
Sense of orientation	O	O	O	O	O	Br. Adam, 1985
Sense of smell	O	O	O	O	O	Br. Adam, 1985
Flying strength	O	O	O	O	O	Br. Adam, 1985
Flight activity	O	O	O			Wilde *et al*, 2003
Flight activity, low temp.	O	O				Büchler, 1988
SCORES:	18	11	6	5	6	
NORMALISED SCORES:	**3.00,**	**2.20,**	**1.20,**	**1.25,**	**1.05**	

O: Best (Score 3); O: Intermediate (Score 2); O: Poorest (Score 1)

Fig. 24. Comparative phenotypes relating to flight and foraging of five honey bee variants. (Adapted from Guerriat, 2013.)

The honey harvest

I mentioned above that my local dark bees produced far superior honey when placed alongside a rape field in flower, compared to striped bees that I think were probably Buckfast. Following that discovery, I replaced the striped queens with uniformly dark ones and the standard of all my honey improved and has remained superb.

In 2002, a standstill order was placed on bees all over mainland Britain, because of an outbreak of Foot and Mouth disease among cloven-hoofed farm animals.

The order meant that we could not take our bees to the heather moors in August and instead, my bees filled their brood boxes with ivy honey from the trees around my home. The following spring, they would not build up, though the cause was not obvious, and it was only several years later that the probable cause was deduced. The most likely explanation was that the ivy honey had probably set solid as it has a very high glucose content, like oil seed rape and so, although their stores were plentiful, they would probably have become inaccessible to the house bees due to crystallisation. I now try to feed my home bees with sucrose at that time to dilute the glucose of the ivy honey that causes it to set so hard.

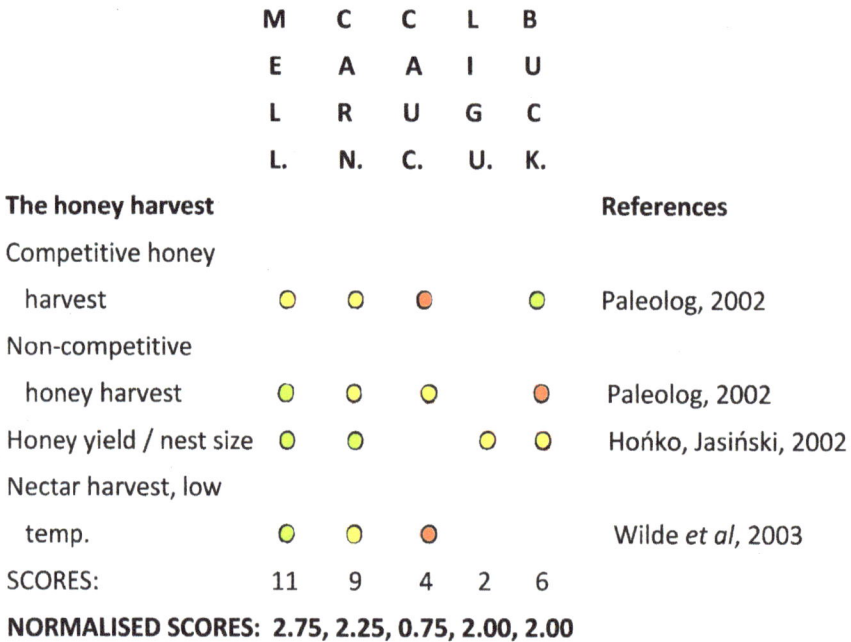

The honey harvest	MEL L.	CARN.	CAUC.	LIGU.	BUCK.	References
Competitive honey harvest	O	O	●		O	Paleolog, 2002
Non-competitive honey harvest	O	O	O		●	Paleolog, 2002
Honey yield / nest size	O	O		O	O	Hońko, Jasiński, 2002
Nectar harvest, low temp.	O	O	●			Wilde *et al*, 2003
SCORES:	11	9	4	2	6	

NORMALISED SCORES: 2.75, 2.25, 0.75, 2.00, 2.00

Fig. 25. Comparative phenotypes relating to the gathering of honey and nectar by the same five honey bee variants. (Adapted from Guerriat, 2013.)

However, I have noticed that the fine flavours of good honey can be destroyed if crystallised honey is heated to melt it. This is often done for bottling or appearance, as the high glucose content of OSR honey in particular causes it to crystallise rapidly, which can look messy. The best solution I have found is to use a bain-marie type water bath, the honey bucket floating in water with its temperature thermostatically held at 51°C by an internal electrical element. At this temperature glucose crystals melt, but not the beeswax comb, which can be chopped up and the liquid honey filtered out. This low a temperature does not destroy the flavour of the honey.

Guerriat's analysis of honey harvests at apiaries across the natural range of *A. m. mellifera* also shows great variation, with *mellifera* again showing a markedly higher score than the others.

Of particular interest are the relative scores for *mellifera* and Buckfast in competitive and non-competitive situations. The "competitive" data relates to a situation intentionally constructed to put the two in direct competition for the same crop, while in the "non-competitive" they were deliberately held apart. When working in isolation *mellifera* scored 3, Buckfast 1, but in the competitive situation the scores were reversed: 1 for *mellifera* and 3 for Buckfast!

Br Adam's Buckfast stock held the World record for the weight of honey collected in a season, but this data shows they were unlikely to have done it on their own. It suggests, in fact, that Buckfast colonies can only gain much of a harvest by robbing the hives of more industrious neighbours!

Some native colonies will work strongly for nectar and water at several degrees Celsius lower than those of other races and earlier in the morning, though showing no difference in warmer weather. Native bees are therefore particularly advantageous to the beekeeper in cool summers. A study in Yorkshire over a 20-year period showed that local natives gained a honey harvest in every summer, in some of which imported bees needed to be fed (John Dews, personal communication).

Pollen collection

As with nectar, pollen harvests are also temperature dependent and native colonies work more strongly than other races at low temperatures, early in the morning and in cool summers.

The comparative behaviours of the races are shown below in Fig.26. Here we see that all aspects of pollen collection, even that at high temperature, is at its highest level among *mellifera* colonies, variable among the others, but particularly low in *caucasica*. A partial explanation for this is that the *mellifera* subspecies is particularly well endowed with "body hair". Their external bristles are more profuse, abdominal hairs are longer and individual "hairs" are branched so that pollen collection is very efficient.

Pollen collection	MELL.	CARN.	CAUC.	LIGU.	BUCK.	References
Pollen collection	O				O	Br. Adam, 80, Cooper, 86
Pollen collecting activity	O	O	O			Wilde *et al*, 2003
Pollen collection, low temp.	O	O	O			Wilde *et al*, 2003
Pollen collection, high temp.	O	O	O			Wilde et al. 2003
Size of pollen loads	O	O				Büchler, 1988
Pollen spectrum pellets	O	O		O		Köppler *et al,* 2007
Pollen spectrum honey	O				O	Delforge, Guerriat, 2012
SCORES:	21	13	3	5	2	
NORMALISED SCORES:	3.00	2.60	1.00	2.50	2.00	

Fig. 26. Comparative phenotypes relating to pollen collection by the same five honey bee variants. (Adapted from Guerriat, 2013.)

A very interesting point relates to the pollen spectrum of the honey, which has a score of 3 for *mellifera*, only 1 for Buckfast, in accordance with my own observations on the honey and pollen collected beside a rape field by dark and stripey bees (see Chapter 3). In that experiment, a considerably greater variety of pollens was collected by the dark bees and the quality of the honey was also much finer with the dark bee colony, in accordance with it having been collected from many flower sources, not just from OSR.

Acclimatisation

One of the most treasured characteristics of the native honey bee is its winter hardiness and ability to overwinter on sparse stores. In part, this is due to its ability to survive in relatively small communities, in part to its ability to survive on its fat stores derived from eating pollen at the end of summer and in part to its early cessation of brood rearing in autumn. It is only in exceptionally hard or damp winters, or after a poor summer when bees have gathered little pollen, or gathered it later than usual, that serious losses occur among native bees. (See discussion relevant to summer and winter temperatures Chapter 3.)

Acclimatisation	MEL.L.	CARN.	CAUC.	LIGU.	BUCK.	References
Coping with bad Weather	O	O	O	O	O	Br Adam, 1985
Overwintering success	O	O	O	O	O	Br Adam, 1980
Winter food use, Finland	O	O		O	O	Hońko, Jasińsk,i, 2002
Winter mortality, Finland	O	O		O	O	Hońko, Jasińsk,i, 2002
Winter mortality, France	O	O	O	O	O	FranceAgriMer, 2012
Longevity	O	O	O	O	O	Br Adam, 1985
SCORES:	**18**	**13**	**6**	**10**	**12**	

NORMALISED SCORES: 3.00, 1.83, 1.50, 1.67, 2.00

	MEL.L.	CARN.	CAUC.	LIGU.	BUCK.
TOTAL SCORES:	68	46	19	22	26
No. of characters:	23	21	15	13	14

NORMALISED SCORES: 2.96, 2.20, 1.27, 1.70, 1.86
PERCENTAGE SCORES: 99% 77% 42% ,57% ,62%

Fig. 27. Comparative phenotypes relating to acclimatisation by the same five honey bee variants and overall figures. (Adapted from Guerriat, 2013.)

Figure 27 shows the relative capacity of the four races and Buckfast to survive winter conditions within the natural geographical range of *A. m. mellifera*.

As expected, *mellifera* was outstandingly the most successful, with a score of 3 for every category, earning an unbeatable normalised score of 3.00 overall. All three other races gave the lowest or intermediate scores for every assessed category except "*Winter mortality in Finland*" for which (except for *caucasica*, which was not assessed) all also scored 3.

In interpreting the data it is useful to refer to the discussion on the natural geographical distribution of the racial subspecies (Fig. 14b), where it is stressed that the natural distribution of the races relates to *summer* temperatures, not those of winter. The implication is that all across Europe, excluding Finland, summer conditions are too poor for adequate collection of stores by exotic honey bee stocks, especially of pollen, by all except *mellifera*, whereas in Finland conditions would seem to be good. It could be however, that the Finnish team (Hońko and Jasiński) took special precautions to ensure their bees of all races were given adequate supplementary summer feeding and this extra attention ensured their survival. The data on "*Winter food use, Finland*" shows that the other races needed a lot of food, as compared to *mellifera*, which was presumably more efficient in its food use.

Whatever the causes, Table 27 makes the point that exotic species need special attention to get them through the toughest time of year. Obedience to my recommendation to "let natural selection rule" would help eliminate the exotics – and that elimination would indirectly improve the local gene pool!

Overview

Figure 27 includes a compilation of the data summarised as total normalised scores of **2.96 for *mellifera*, 2.20 for *carnica*, 1.27 for *caucasica*, 1.70 for *ligustica* and 1.86 for Buckfast,** out of 3. These are also expressed as the percentages: **99%, 73%, 42%, 57% and 62%** respectively, emphasising the point that *A. m. mellifera* is by far the fittest subspecies *in its own natural range*.

Although this point is predicably obvious to biologists and intelligent beekeepers it has not prevented our native species being taken close to extinction by the very irresponsible practices of importing exotic subspecies and propagation of the artificial multi-hybrid Buckfast.

A Scottish honey farmer told me the reason he does not use native bees is that their honey yields are too low. He said he needs a crop of 250lbs of honey per hive, which he can get (in Scotland) by force-feeding imported bees with sugar syrup. I have never aimed for large honey crops, and the most I have achieved was an average of around 150lbs, from four hives. Each hive had both a young and an old queen at peak nectar flow, the young one alone in the combined colony later at the heather, and the same young one in winter. There was no artificial feeding at any time.

Inherited characters according to caste and sex.

Since the various characters of a colony are expressed either by the queen, her workers, or her drones, it follows that breeding for a character, or the performance of a particular task, should address the appropriate caste or sex. This section segregates the tasks according to caste and sex, as defined by Cooper (1986).

Characters conferred by the workers
Capacity for being subdued.
Body colour.
Following.
Jumpiness.
Use of propolis.
Comb capping and cleaning.
Queen cell number.
Inclination to swarm.

Characters conferred partly by workers, partly by the queen
Size of the broodnest.
Number of drones.
Tendency to ball cross-bred queens.

Characters conferred predominantly by the queen

Longevity of the queen.

Aggressiveness to rival queens.

Disposition to pipe during swarming.

Stinginess due to deficiency of queen substance.

Characters conferred by the drone's genome

Colour and shape of drones.

Flight temperature activity of drones.

Drifting of drones.

Disposition towards AVM, as compared to drone assembly mating.

Inherited bad temper.

Chapter 6

REPRODUCTION

Proneness to swarm

The ultimate aim in establishing a conservation apiary is to ensure colonies can look after themselves and remain vigorous and healthy without much attention from a beekeeper. I have in mind the need for this to continue for a thousand years! That means they should swarm naturally, but only at a fairly low frequency. There should be a few empty hives on site to catch swarms, but a beekeeper may still be needed to check things are in hand.

Our summer season is so short in Northumberland that colonies resulting from swarming are often unable to build up to strength before winter. I therefore try to re-hive the prime swarm with the laying queen, set up two or three nucs from surplus natural queen cells and leave one or two good queen cells in the swarmed colony. As the summer progresses, I identify the best nucs, then at the end of summer, remove the poorer queens and unite two or three colonies under each of the best queens. That way my new colonies go into winter with strong workforces led by the best of the new queens.

Fortunately, proneness to swarm is also expressed in the number of queen cells produced when a colony prepares to swarm, so I don't allow daughter queens to be raised from colonies where there are more than 6 queen cells and that probably helps keep my swarming frequency down.

Most of my stocks swarm every one or two years, but it depends on the amount of nectar coming in. I could possibly move some hives into the second- year category if I set them "brood-and-a-half", that is with brood in both a deep and a shallow box, with the shallow on top and no queen excluder between them. But I don't like this conformation for ordinary use, as it takes twice as long as a single box to check through for queen cells. I know a Yorkshire family that swear by brood-and-a-half, as when their bees prepare to swarm, they say there are always queen cells at the interface. It is then an easy matter to check, as you just need to lift one side of the shallow box to see them protruding down. I have tried this more than once but find it unreliable with my bees.

Non-prolificacy

Much of this book is informed by the thoughts and words of Beowulf Cooper (1986) the traditional authority on the native British honey bee, who stressed that the subspecies' most important single characteristic is non-prolificacy. What sounds like a negative does in fact confer major advantages in economy and scale and enables colony survival in times of shortage.

Non-prolific bees are easier to handle and often just as productive in terms of honey production as much more prolific strains. For example, Italian bees will accelerate laying and brood rearing in response to a bountiful summer harvest, but omit to reduce brood productivity when the harvest is over. The bumper harvest can then be entirely consumed by the multitude of workers. Commercial beekeepers cope with this by removing honey supers at their peak, then making up the colony's loss by feeding copious quantities of sugar syrup. By contrast, British native queens being sensitive to nectar income, slow their egg laying, so that the hive population declines with the forage. The honey harvest is thus saved and the bees' diet remains rich in its essential nutrients (See "Feeding bees", Chapter 3).

Longevity

Long-lived queens will usually beget long-lived workers and, according to Cooper, some *A.m. mellifera* queens can go on laying into their 8th, 9th or even 10th year. Some long-lived queens are slow to build up in their first year, but it can be worth keeping them, as they may show their strengths later. An advantage of having long-lived workers is that extension of their life tends to occur in the later stages of maturation. This means that you get a higher proportion of skilled foragers in the workforce and hence more honey per bee.

Supersedure.

Supersedure (spelled with an "s" not a "c") refers to the production of an extra young queen, whose task it is to supplement the activities of her failing mother. Supersedure queen cells are typically only one or two in number and usually are positioned midway up a brood frame, about one third of the way from one end. Their construction is generally a comment on poor laying performance by the queen, that can be due to ageing or follow non-lethal injury to her, or other cause of reduction in egg production.

After mating, a young supersedure queen returns to her hive of origin without swarming having taken place. She then, without aggression between the two, lays in the same hive as her mother and this continues for some time before she takes over as sole egg producer.

Supersedure does not replace swarming, but in the colder regions of Northern England, it provides a safe way for the colony to reproduce without the serious risks associated with moving house. It is definitely a trait to be retained in your stocks if possible.

Brood breaks

Native British queens cease laying for several reasons. It occurs routinely preparatory to winter and continues for two months or more (see Fig. 28). It also occurs when there is a dearth of nectar and just before a laying queen swarms.

When the queen stops laying for any reason the gap in brood production can have unexpected benefits. Brood diseases such as chalkbrood and European or American foulbrood suffer a check, which can be beneficial for the bees, especially if they practise hygienic behaviour and clean out contaminated brood cells. It is especially beneficial in relation to varroa infestations because the mites then lose their safe sanctuary away from allogroomer bees, which can kill mites with their mandibles (see Appendix).

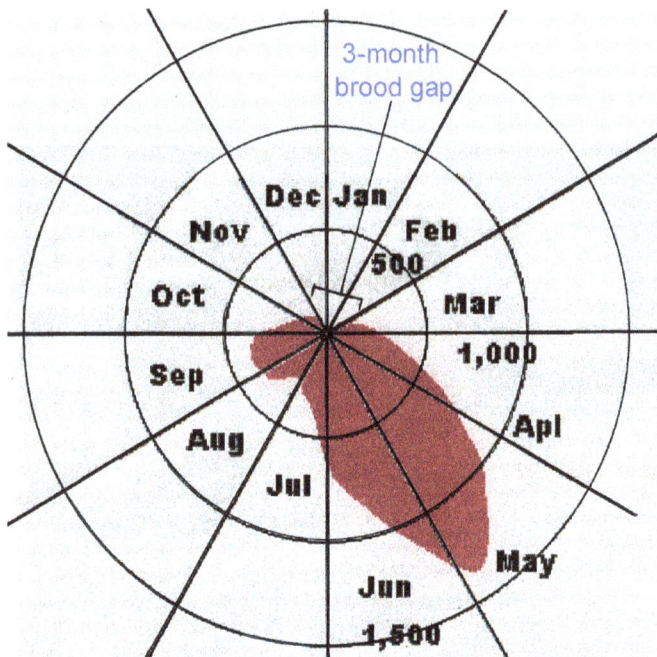

Fig.28. Annual variation in broodnest size. (Note the major break in brood rearing roughly from the beginning of December to the end of February, when adult varroa mites are particularly vulnerable to grooming bees (Cushman, D. website)

Culling bees

Nature's favourite means of "improvement" is by culling out weak stocks, so that average performance is raised. Although upsetting to perform, killing an undesirable colony such as one that is overly aggressive can be very effective, so all bee conservationists should be prepared occasionally to follow Nature's example. Indeed Cooper (1986) advises that high priority should be accorded to culling.

Various aspects of culling can be carried out by active killing of a whole colony, starvation, relocation, dequeening with or without also removing the drones, and destruction and replacement of drones.

A honey bee colony is killed rapidly by exposure to petrol vapour, or by shaking bees onto frost or snow on the ground. These approaches are also useful for acquisition of comb or food for more productive stock, to do away with laying workers, dispose of queenless nucs, eliminate vicious bees, remove swarms from inaccessible cavities like chimneys, and disposal of diseased stock. For unwanted swarms in chimneys, some beekeepers use "fireworks" that emit insecticide as used in tomato whitefly infestations. These are illegal for this purpose and dangerous not only to humans, but also particularly to fish.

Starvation can be used for eliminating (foreign) colonies that are undesirably over-prolific during forage dearths.

Relocation is particularly useful for controlling mating by transfer either of the virgin queen stock or her desired mating partners to a new site. It is valuable as a way to test non-local genomes in your apiary without major commitment to the untested genome.

Dequeening of a poor colony and uniting is a useful strategy for creating strong colonies to catch a honey flow, to reduce the number of colonies to be fed before winter, to contribute to the spring build up, or for propagation or pollination duties. Its use is common in heather areas for creating a massive work force ahead of the nectar flow in August.

Drone free stock can be achieved by removing drone brood, or by taking workers from the supers. It can be useful to ensure a bad character trait such as vicious behaviour is not spread by gene flow to other lines. The converse, of propagating drones of gentle or otherwise beneficial stocks is much to be recommended. I routinely place a frame of drone foundation or drawn drone comb in each of my good stocks to ensure the most desirable drone population is maintained at a high level.

How many drones are enough?

A strong drone assembly contains a great excess of fertile drones, ensuring that the fittest of very many get to mate with the virgin queen. But the important point is not so much the number of drones, but the variety of their genomes. There are two issues here. The first is the requirement for many versions (i.e. alleles) of the multitude of genes responsible for local adaptation and survival through ancestral hazards. That is their secret weapon against all manner of adversities. The second is the need for many alleles of the gene that defines sex, the *complementary sex determiner* gene, or *csd*, and this for a very different reason. Insufficient genetic variety in both respects is highly detrimental, so it is important that several unrelated drone producing colonies are provided for mating. I suggest representatives of a minimum of eight different queen lines, all of good native type, in addition to one or more nucs containing the virgin queen(s; see "Inbreeding" below).

In the Northern hemisphere, honey bees are under serious threat, particularly from varroa, agricultural chemicals and historical mismanagement. But *in extremis*, the limiting factor is none of these, but a genetic quirk of bee biology that demands heterozygosity (i.e. two different alleles) of the *csd* gene. (see Chapter 4) This is necessary not only for definition of female sexuality, but also to avoid the hopeless outcome of homozygosity. Because if a virgin queen mates with a close relative and her ova get fertilised by sperm carrying the same allele of the *csd* gene, the outcome is homozygous larvae that develop into what we know as "diploid drones". These are non-functional males that are recognised and destroyed by the house bees on emergence from the egg. This lowers the effective fertility of the queen, reduces honey yield and most importantly, places overwinter survival of the colony at considerable risk.

In very close inbreeding, for example, of a queen with her brothers, as many as 50% of her fertilised ova would be homozygous for *csd*, and the queen's effective fertility is reduced from 100 to 50% (see Fig. 29). The beekeeper can tell when this has happened by the number of empty brood cells in her brood combs and by the probable loss of that colony the following winter.

Ova genotypes

	a1	a2
a1	a1a1 DD	a1a2
a2	a2a1	a2 a2 DD

Sperm genotypes

Fig. 29. Punnet square showing the outcome of mating between the daughters and sons of an a1/a2 queen. Two kinds of homozygotes are formed, both of which become diploid drones and are eliminated by the house bees.

A mature queen produces ova that can be fertilised by sperm and turn into females that, by modification of their feeding, can be caused to develop as either worker bees or queens. If her ova are not fertilised, they develop instead into drones. In every cell of her own body, the mature queen has just two different *csd* alleles and each of her ova, and each drone she produces carries either one or the other of her *csd* alleles. That is, she releases two and only two types of *csd* allele into the surrounding population, carried in her eggs and the sperm of her sons.

Her spermatheca stores the sperm she received at mating. These go one at a time into ova as they are laid and initiate their development as females, but they are not otherwise released into the population. So, every queen is herself heterozygous for *csd* and contributes copies of both her *csd* alleles, but no others, to the wider population, through her drones.

If the wider population contributing to the drone assembly has many colonies releasing drones with many different *csd* alleles, the frequency of those alleles tends to equilibrate in more or less equal proportions. So that, for example, if there are twenty *csd* alleles altogether, each would tend to be present at a frequency of around 100 / 20 = 5%. If there are 10 alleles each would be at around 10%. It can be seen that the frequency of *csd* homozygotes and hence diploid drones in a colony would increase or decrease in relation to the population frequency of those alleles that are also present in the mother. If there are many different alleles altogether, their individual frequencies are low, and if mating is at random, the frequency of diploid drones would also be low. On the other hand, if there are few *csd* alleles *in toto*, the frequency of each would be high and that of diploid drones would also be high. (Fig. 19).

In Chapter 4, we learnt of Tarpy and Page's (2002) observations that for all colonies to survive the winter the brood viability needs to be 72% or better. (The values of 75% for viable and 25% for inviable are used by Ruttner (1988) in his analysis of inbreeding – see below). Even with a quarter of the brood nest non-functional, the colonies survived, but when numbers of diploid drones pushed the proportion of empty brood cells above 28%, 6 out of 10 colonies succumbed to the winter cold. The probable explanation is that colonies with low brood viability have difficulty generating sufficient communal heat to ensure survival in timber hives.

What this means is that high levels of *csd* polymorphism in the drone assembly are necessary to promote temperature maintenance and winter survival of colonies led by the queens that mated there. It is not generally recognised, but ***maintenance of high levels of polymorphism of the csd gene must therefore be considered one of our highest priorities in honeybee conservation.***

Mathematical analysis of Tarpy and Page's observations revealed that the limiting diploid drone density of 28% arises, in freely interbreeding populations, where there are 6 *csd* alleles, that are each at a frequency of 17%.

This analysis did not incorporate other factors known to reduce brood viability, such as age of the queen, presence of chalkbrood, extreme weather conditions, or inappropriate timing of brood nest development, etc. So, it is wise actually to aim for a larger number of drone *csd* types. I suggest 10 or more, which would require at least 5 genetically unrelated queens. I try to maintain representatives of half a dozen native or near-native queen lines set up from swarms caught at different locations within 50 miles of my home apiary. An extra frame of drone comb in each of these ensures massive and varied drone production, without significant reduction in the honey harvest.

Inbreeding

Inbreeding means the mating of individuals that are related to each other by ancestry. Among honey bees, the usual situation is when a young queen mates with her brothers or first cousins. This is most common among small populations, especially pioneer populations, or when the weather is too poor to allow mating flights to drone assemblies at a distance from the hive. Typically drones gather in mating assemblies fairly close to their own or other hives, as they need to refuel with honey many times a day, whereas queens may indulge in only one or a very small number of mating flights or often over-fly local drone assemblies. By the latter strategy queens usually tend to be successful at finding an adequate number of unrelated partners. An exception to this is during bad flying weather when native queens may mate close to home, typically at hedge-top level instead of 30 feet up in the air. This is known as "apiary vicinity mating" or AVM and is one of the adaptive strategies of the native subspecies.

One consequence of inbreeding is "inbreeding depression" perhaps revealed most markedly by reduced honey yield. Other indications are sluggish build up, poor colony strength and high winter losses. Theoretically, another outcome would be recessive disease, i.e. disease caused by homozygosity of harmful recessive alleles. But this is probably rare in honey bees because of the haploidy of drones, which could cause recessive alleles to be expressed and if detrimental, eliminated by loss of the drone.

As explained in Chapter 4, diploid drones arise as a consequence of homo-zygosity of alleles of the *complementary sex determining gene, csd*. This can occur when populations have low levels of polymorphism of *csd*, or more commonly due to inbreeding. Indeed, failure of worker bee larvae in an otherwise healthy brood nest, with appearance of "pepper pot" or "shot" brood is usually taken as evidence that inbreeding has taken place, although patchy brood can also arise from EFB.

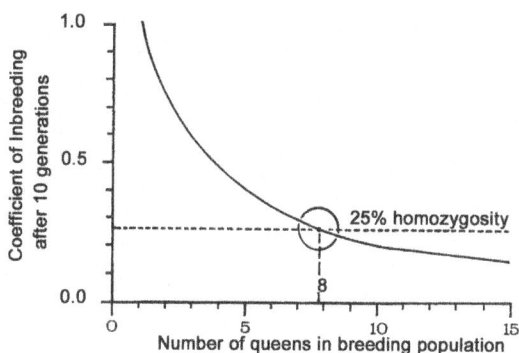

Fig. 30. The degree of inbreeding after 10 generations of selection of different numbers of queens per generation. The critical value of 25% inbreeding is reached when only 8 breeding queens per generation are selected. (Redrawn from Moritz, 1984.)

Figure 30 resurrects the concept mentioned in Chapter 4 in relation to *csd* polymorphism and overwintering success, where we saw that survival of a colony is virtually guaranteed so long as fewer than 25% of the brood is inviable (i.e. and at least 75% is viable). Figure 30 is taken from Ruttner (1988) and answers the question *"how many breeding queens do there need to be to keep inbreeding below the critical value of 25%?"* At first glance it looks very similar to Fig.19, dealing with the number of *csd* alleles in the population, but this second graph deals with a different question and from a different viewpoint.

What Fig. 30 shows is the relationship between the mathematically derived "coefficient of inbreeding" over 10 generations and the number of queens in a closed group. In this case, the coefficient was derived by computer modelling and indicates the probability of two homozygous alleles present in an offspring of the 10th generation being identical by descent from their common ancestor.

This is important in the context of the *csd* gene as its homozygosity would results in a diploid drone, its death soon after emergence and if repeated in many larvae the possible extinction of the colony. Fig. 30 can be understood as showing the relationship between formation of diploid drones due to convergent descent of one *csd* allele and the number of individual queen lines that have crossed with one another in those 10 generations.

The 25% rule (or 28% - see "How many drones are enough?" above) applies here and reveals that the cut-off between 100% over-winter survival of colonies and a much lower figure is at 8 queens: fewer than 8 queens in the closed breeding group and the risk is high, more than 8 queens and the risk is very low, again due to the effect of large numbers of diploid drones on colony survival.

So, the" take-home message" is that if you have a potential inbreeding situation, the drone assemblies where your virgin queens mate must be supplied by drones from 8 or more queens or your colonies are unlikely to survive the winter. Fewer than 8 drone-producer queens and the probability of homozygosity of *csd* due to convergent descent is so high that your colonies would be below the tipping point mentioned in Chapter 4, with a high probability of their dying from the cold during the next winter. A measure you could try to save them would be to equip the hive housing the queen with better insulation and supplement the colony with additional worker bees.

Chapter 7

DISEASE AND DISEASE RESISTANCE

Resistance to Varroa

The varroa mite (*Varroa destructor*) is now considered the deadliest enemy the honeybee has, although it is rarely the mites *per se* that are the harmful, but the viruses they carry, such as *deformed wing disease* (*dwd*) virus. Moreover, some native stocks can mount considerable resistance to the mite.

The term "Varroa Sensitive Hygiene", or VSH, is generally applied to one of several aspects of resistance that involves uncapping infested brood cells and removing and killing the mite(s) inside. Some British natives may have this ability, but my Northumberland bees, John Dews' Fountain's Abbey bees (now in Whitby) and Rodger Dewhurst's native stocks in Cornwall defend themselves mainly by killing mites with their mandibles, in the behaviour known as "grooming", or "allogrooming" (meaning "grooming another"; see Appendix). This and any other varroa resistant behaviour should certainly be conserved in your stocks if it is suspected.

A varroa trap?

While running a practical class, a student asked me what did I do to guard against Varroa mites? I said I did nothing, but let the bees sort it out for themselves. I told the class they needn't worry about varroa in my hives, as they would be unlikely to see any. Then I came across three sealed queen cells, which was unexpected as it was rather late in the summer. I did not want more queens at that time, so I cut the cells off with my hive tool. Two looked normal on the outside and the queens inside also looked normal, but the third was very different. Its outside surface was smooth, with thick wax, so that it reminded me of a Singer sewing machine shuttle. I broke it open. The queen prepupa looked rather bloated, but otherwise appeared essentially normal, but on her body were 14 adult varroa mites! (Fig. 31).

Fig. 31. A possible varroa trap. The reinforced queen cell plus prepupal queen and 14 adult mites.

Varroa mites do not normally enter queen cells, as the royal jelly they contain is repulsive to them, unlike that of worker and drone cells. Chemical analysis and laboratory experiments both indicate a relatively high concentration of octanoic acid to be a major factor in this repulsion (Nazzi *et al*, 2009). Octanoic acid is also known as caprylic acid, due to its presence in goats' (and incidentally also human) milk, from the Latin *"capra"*, a goat.

I wonder if this was an entirely chance occurrence, or whether these house bees had set a trap for the mites. That the surface of this cell had apparently been reinforced with extra wax implies the bees knew the mites were inside and were determined not to let them out. This hive was not from the same queen line as Colony JB5 (see Appendix).

Chalkbrood

Honey bee diseases of all kinds are a rarity in Northumberland, except for chalkbrood and it is generally advised that the best cure for this is requeening. Cooper (1986) suggested that fairly mild outbreaks occurring after very cold nights may be the result of fluctuating broodnest temperatures. This is a characteristic of *A. m. m.* But some strains are such good house cleaners that they prevent its building up and eject diseased larvae and pupae before they can produce infective spores.

Cooper believed that susceptibility is undoubtedly inherited and inbreeding susceptible stocks seems to worsen the tendency. Stocks very susceptible to chalkbrood should therefore not be used as breeding material.

Nosema

Nosema apis is a fungal protozoan that causes dysentery. It can be eliminated in a couple of seasons by good housekeeping strains and by bees that practise comb varnishing. Excitable bees prone to excrete in the hive during manipulations or when shut in tend to be susceptible.

Foul brood

Two types of foul brood occur in Britain, American (AFB) and European (EFB). An easy way to remember a major distinction between them is that **E**uropean occurs **E**arly, before brood capping has occurred, **A**merican occurs **A**fter the brood has been capped.

There are several types of inherited resistance to AFB, but British regulations dictate that all cases must be reported and destroyed under supervision. EFB, known on the Continent as "sour brood", is less serious, but it is wise to assume all strains of honey bee are susceptible. EFB responds to antibiotic treatment, but like AFB, it is also a notifiable disease.

Wax moth

Wax moths tend to infest stored comb and can be a nuisance by spoiling the appearance of cut comb by burrowing under the cappings.

Some bee strains are tolerant of attacks by small wax moths and keeping colonies strong and in secure hives can also help eliminate them. Empty comb is ignored by the moths if hung in the open air, as in an airy shed, which is useful to know with regard to winter comb storage.

Greater wax moth gets spread by fishermen emptying their bait tins at the end of the expedition, as the larvae are sold as fishing bait.

Acarine disease

Acarine disease is caused by a mite infestation of the tracheae or breathing tubes and used to be widespread in Britain. According to Cooper, most stocks are moderately resistant to acarine, but a few are highly susceptible, some of which are the outcome of crossings between the moderately susceptible.

Acarine mites are very small compared to Varroa mites and are probably vectors for chronic bee paralysis virus (CBPV), which was the probable cause of the devastating epidemic known as "Isle of Wight Disease".

Isle of Wight Disease, an ongoing puzzle

During the first two decades of the 20th century, the British Isles were swept by a honeybee epidemic, the first reports of which were from the Isle of Wight. It became known as Isle of Wight Disease and it devastated honey bee stocks all over Britain. According to widely quoted official estimates 90% of British honeybee stocks were wiped out, including all native British *Apis mellifera mellifera* (Adam, 1968, see below). The Buckfast strain was created by Brother Adam at Buckfast Abbey as a response to this epidemic. However, some native stocks were known to have survived in the care of beekeepers and in the 1970s, the first post-epidemic native feral colonies were found by William Bielby at Fountains Abbey in Yorkshire.

In the British Bee Journal of February 8th 1906, there appeared a short report with the heading *"Bee paralysis: is the cause known?"* It went on: *"During a bright, sunny day, recently, the bees coming out of the hive dropped onto the grass by the dozen and seemed quite powerless, so far as using their wings".* The abdomens of some were distended and the wings of many twisted back, giving the appearance of being dislocated. *"They kept crawling up the blades of grass and onto the alighting board, only to tumble off again…..and when the day became colder, the bees collected in little bunches of a dozen or so in each, and soon perished".* Another hive became affected with paralysis, developed dysentery, and rapidly dwindled away.

Infective Bee Paralysis?

The author, a Mr HM Cooper, predicted that if the disease crossed from his apiary on the Isle of Wight to the mainland, *"it would probably mean the ruin of the bee industry, so far as England is concerned"* (Cooper, 1906). The editor of the BBJ poo-pooed this alarmism, with disparaging remarks about Cooper and expressed his view that it was "*nothing more than paralysis*", in support of which he quoted a lengthy passage from "*The ABC of Bee Culture*".

A year later beekeepers on the island were getting worried and a widely respected John Silver was invited to advise. He made the first independent confirmation of the extent and gravity of the outbreak and the BBJ editor at last began to take it seriously. He reprinted an extract from the Annual Report of the Hants and Isle of Wight, BKA entitled: "*Bee Epidemic. Isle of Wight Scourge*" (BBJ, 1907). It included the statement: "*a new - or at all events not understood - disease has spread east and west, north and south through the island*". Paralysis was dismissed because bees with paralysis are frequently reported as being black, hairless and shiny, while the IOWD casualties were not (see below). Fifty years later, Brother Adam wrote that by 1907 "*it was fully accepted that paralysis had nothing to do with the outbreak* (Adam, 1968).

Nosema?

"Spring dwindling" of colonies was then a problem, but in 1907 Prof. E Zander deduced that its typical cause is the unicellular parasitic gut fungus, *Nosema apis* (Zander, 1907). The reported signs of nosema and IOWD were rather similar and a team of scientists engaged by the Board of Agriculture initially concluded that *Nosema apis* was responsible for IOWD. Nosema indirectly causes dysentery, although almost exclusively in spring. Also, with nosema infestations there is typically no mass crawling, or if it occurs it is only in June or early July. A major distinction was that queens are particularly susceptible to nosema and often among its first victims, but in IOWD the queen usually survived to the last. In 1916, a Dr J. Anderson stressed the dissimilarities and non-identity of the two diseases (Anderson, 1916).

Because of the diagnostic confusion and the complications related to the new use of moveable frame hives, the management of which is very different from that of skeps, there were many failures of management, but the tendency was to blame human errors on IOWD. The apiary at Buckfast Abbey contracted the disease in 2015, and it became clear to Br Adam that something new was killing their bees, and not nosema, AFB, EFB, sacbrood, chalkbrood, septicaemia or poisoning (Adam, 1968).

Acarine disease?

Diagnosis was proving unusually problematical, then in 1917 Dr John Rennie of the North of Scotland College of Agriculture, made a breakthrough. After much experimentation he concluded *"I have found that at least 99% of stocks reported to me as failed or failing from what has been popularly termed "Isle of Wight disease" harbour the parasite Tarsonemus woodi"*. This organism was later renamed *"Acarapis woodi Rennie"* and popularly called the acarine or tracheal mite.

The acarine mite is a parasite of bees that lives within the tracheae of the thorax and head, where it feeds on the host haemolymph. Adult female mites invade the host through the spiracles where they lay on average just under one egg per day and as many as 21 in a lifetime. The only time the mites live outside the host is when young females transfer to new hosts (Pettis and Wilson, 1996). In warm climates, honey bees can sustain heavy infestations of acarine mites with little impact on their health, but in colder climates, their impact is much greater. Tracheal mites cause adult bees to leave the hive, but affect their ability to fly and defecate, leading to large losses of young adult bees in winter and spring (Webster and Delaplane, 2001). At the time the available evidence seemed to give support to Rennie's diagnosis of acarine disease being the cause.

Acarine mites are microscopic and infested bees can be asymptomatic, thus presence of disease is easy to miss. A typical sign is a weak colony in early spring, where the returning foragers do not make it all the way back to the hive. Many then must crawl the last few centimetres to the hive entrance, resulting in large numbers of bees crawling in front of the hive. Pathogenic effects depend on the number of mites within the tracheae, where they are responsible for both physical injury and physiological upsets. Wing dislocation sometimes occurs and seems to be associated with gatherings of mites at the wing roots.

Coincidence of acarine mites and IOWD was however by no means 100% and experimentation by Lesley Bailey in the 1950s and 60s at Rothamsted Experimental Station in Hertfordshire cast further doubt on acarine being the cause. Even back in 1923, Rennie had admitted several maladies with analogous symptoms had initially been included under the original designation of "Isle of Wight Disease", which can only have caused confusion.

Deliberate infestation with *Acarapis* led to only a small proportion of heavy infestations and this tended to depend on whether the previous or current season was poor or good by British standards (Bailey, 1961). Furthermore, the proportion of crawling bees with IOWD that carried mites in their tracheae was nowhere near 100% (Bailey & Lee, 1959).

Although individual bees have their lives shortened by acarine infestation the effect is only slight. There is also a seasonal effect; even in very heavily infested colonies, disease is not usually visible in summer (Bailey, 1961), though mortality and mite infestation correlated more highly in winter (Bailey, 1958; Bailey and Lee, 1959).

In one much publicised presentation to the Central Association of Beekeepers, Bailey (1963) referred to IOWD as a myth, which he defined as: "*a primitive explanation of a natural phenomenon*". He mentioned Malden a professional investigator, who had earlier pointed out that the swollen intestines of bees with IOWD resembled those of healthy bees that had merely been confined to their hives for a few days (Malden, 1909). Malden examined the anatomy of diseased bees minutely, including their tracheae and air sacs, but found no mites in the tracheae, only more bacteria in the guts of diseased bees than healthy ones, and he was unable to show these were pathogenic.

Bailey (1963) mentioned that Bullamore (1922) had described bees prevented from flying as having symptoms described as "crawling" and with bowel distension that was indistinguishable from that in IOWD. He also mentioned large scale losses in recent years in countries other than Britain, particularly Australia and South America. These had symptoms exactly like IOWD, although no known parasite was found to be present.

Bailey claimed the results quoted as proof of causation by Rennie, White and Harvey (1921) did not actually support the idea that *Acarapis woodi* had caused the disease. On the contrary, he claimed the mite was widespread, and although it occurred in all the colonies the authors believed to have IOWD, it was also present in many normal colonies. The report of Rennie *et al* (1921) showed that many bees from both diseased and healthy stocks behaved normally and many flew normally, even when infested with mites, some with pronounced blackening and hardening of their infected tracheae.

Chronic Bee Paralysis Virus?

In the early 1960s, scientists at Rothamsted turned their attention to viruses, and slowly a sensible picture began to emerge. The causative agent was actually confirmed in 1963, when Bailey and colleagues isolated and characterised Chronic Bee Paralysis Virus (CBPV), but the picture was still complicated by the probability that viruses may be carried between hives and between individual bees by acarine mites. Another great complication was that CBPV has two distinct syndromes! (Ball and Bailey, 1997; Bailey and Ball, 1971).

You will remember that the editor of the **BBJ** initially diagnosed paralysis as the cause of IOWD, but his interpretation was dismissed. That was probably because comparisons may have been made with the wrong syndrome. In Syndrome 2 affected bees are able to fly, but are almost hairless. They appear small and dark, are nibbled by other bees and hindered at the hive entrance by the guards. A few days after infection, trembling begins, then they lose their ability to fly and soon die.

In Syndrome 1 the bees cannot fly from the start and often crawl on the ground and up plant stems, trembling their wings and body. There they huddle together and on the top bars of the hive, sometimes in very large numbers. They may have bloated abdomens due to distension of the honey sac, which accelerates the onset of dysentery. In Syndrome 1, the wings are typically partially spread or dislocated, and sick individuals die within a few days of the onset of symptoms. Severely affected Syndrome 1 colonies suddenly collapse, particularly at the height of summer, typically leaving the queen with a few workers on neglected combs. All these symptoms of syndrome 1 are identical to those attributed to "Isle of Wight Disease".

Both *Acarapis woodi* and CBPV are contagiously transmitted and tend to increase in prevalence under similar circumstances (Bailey et al, 1983). It is Syndrome 1 that is most widespread in Britain. So there is every excuse for confusion, but *the presently accepted explanation is that IOWD was caused not by acarine mites, but by CBPV expressed as Syndrome 1.*

The Buckfast bee

When IOWD arrived in Devon there were several honeybee subspecies and strains in the care of the monks at Buckfast Abbey. The disease hit them hard and most colonies died, but the survivors notably included bees of the Ligurian ecotype of the Italian *Apis mellifera ligustica*. And, although all the native British *A. m mellifera* at the Abbey died, first generation crosses between them and Ligurian bees survived. The Buckfast IOWD-resistant strain was set up from these survivors, with additions from several other strains collected by Adam on visits to remote areas of Europe where he expected local natives might still exist in pure form.

Adam previously fully accepted Rennie's interpretation of acarine mites being causative of IOWD although latterly Bailey dismissed the idea in favour of CBPV. In the context of the acarine causation theory, the resistance of the Ligurian bees was ascribed to stiffer bristles across the respiratory spiracles preventing mites from entering the bees' tracheae. Normally these bristles are initially soft and flexible, but they stiffen over the first five days after emergence from the pupa, so that mites can pass through the spiracles only before then.

Both Rennie and Bailey & Ball (1991) wrote that acarine mites were always or almost always present in bees affected by IOWD, although they differed in their interpretations of the mite's significance. An explanation fitting both interpretations would be *that acarine mites can, or may not, carry viruses, but entry of CBPD viruses into the bees' bodies is facilitated by the entry of mites acting as their vector*, like Varroa mites acting as vectors for the virus that causes Deformed Wing Disease. That idea would seem to resolve the apparent discrepancy.

Could IOWD come back? Unauthenticated reports imply that hotspots of CBPV remain in Britain and could cause trouble in future. If it does, I hope the above analysis will facilitate its diagnosis.

Chapter 8

RECORD KEEPING AND SECURITY

Records written in the apiary should be brief, but comprehensive on all important issues and intelligible to other readers.

I prefer them to be accessible away from the apiary and handwritten in a book, rather than on cards that can get lost, fall out of order or be blown away. I therefore use a simple notebook, purchased from W.H. Smith and similar to those used by the Dragons of TV's "Dragon's Den". These have a ribbon to allow pages to be easily relocated and an elastic loop fastener so that books remain open on a hive roof in the wind.

Below is an outline of my typical records under three headings, Hive identity, Report and Recommendations.

Hive identity

1   Location or identity of apiary, e.g. "Orchard".

2   General layout and appearance of hives from the front: rough sketch, labelled.

2   Hive number (e.g. "23") in black upper case on a white plastic sheet pinned or stuck to front of brood box.

4   Location of hive in apiary, e.g. "Far East" or "3rd from left".

5   Name, identity, or origin of queen, e.g. "ELLA", "ELLA'S D2" (i.e. Ella's second daughter), "WHALTON NORTH".

6   Colour marking of queen, represented by a coloured drawing pin on the hive front, indicating also the year of her birth, in the order: **W**hite, **Y**ellow, **R**ed, **G**reen, **B**lue for years beginning with 1 or 6, 2 or 7, 3 or 8, 4 or 9, 5 or 0, respectively. I remember this order by the mnemonic: **W**ill **Y**ou **R**ear **G**ood **B**ees?

Report

1   Date of inspection.

2   Temperature and weather.

3   Activity at hive entrance, note if pollen is being taken in (indicating the colony is queenright and brood is being raised).

4   Presence of ginger-banded workers (indicating hybridisation and probable susceptibility to Varroa; see Appendix).

5   Number of frames of brood and/or eggs.

6   Brood pattern – solid or patchy (indicating level of inbreeding).

7   If no brood, presence of polished brood cells (indicating preparation for laying) and whether colony "feels" queenright; or diagnosis of a drone layer (by presence of small drone brood, in place of worker brood).

8   Presence of pollen stores ventral to (i.e. below) the brood, indicating native status (see Fig. 13).

9   Presence of drone brood in Spring in addition to worker brood; note to expect queen cells two weeks after the first drone brood.

10   Sight of queen; check on her paint colour compared to that of marker pin on hive front.

11   Need for super, repair of hive or other requirement.

12   Mood of colony, aggressiveness or over-defensiveness; suitability for breeding, or not.

13   Swarming preparations – note to split colony or set up nuc at identified date.

14   Honey stores if possibly inadequate to last till next inspection (usually only necessary at new or "sparse" locations).

15   Any unusual observations, e.g. "robbing by wasps".

16   Other relevant observations, e.g. "super two-thirds full".

17   Actions taken, e.g. "drone foundation inserted", "wasp traps set up".

Recommendations (indicated with a prominent mark, such as a heavy cross in the notes if urgent):

1   "Feed with sugar syrup".

2   "Feed with Candipolline" (pollen and sugar supplement).

3   Equipment renewal or replacement, e.g. "insulation board needed"; "BB needs repair", "replace entrance block with smaller entrance".

4   Honey harvest – e.g. "super needed", "insert clearer board Tuesday a.m."

5   Intended mating by drones in 4-5 weeks' time: "add drone foundation".

6   Mating of queen(s): "take to (named) apiary".

7   Robbing by wasps; "reduce entrance", "set up wasp traps".

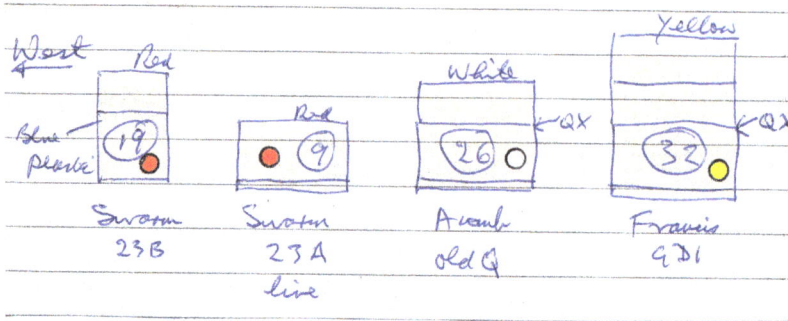

Interpretation: from West to East

Hive 19 is a blue plastic nuc holding the 2nd (B) swarm caught in the Orchard Apiary last season, 2023. The queen is marked Red (born 2023).

Hive 9 is a timber National brood box only, holding the first (A) swarm caught in 2023, which is still alive. The queen is Red (born 2023).

Hive 26 is a timber National holding the old queen of a colony caught in Acomb. It has one super over a queen excluder. The queen is White (born 2021).

Hive 32 is a timber National with a queen excluder and under two supers. The queen is marked Yellow (born 2022). The queen is the first granddaughter of the queen given me by Francis.

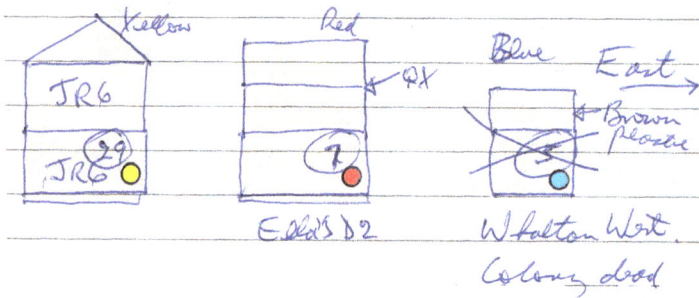

Hive 29 is a timber National with an ornamental roof and holds Colony JR6 (descendent from one given me by John Reay) in two deep boxes. The queen is yellow (born 2022).

Hive 7 has the colony with the second of Ella's daughters on "brood-and-a-half", with one super. The queen is marked Red.

Hive 5 is a brown plastic nuc. It holds a colony caught as a swarm at the West end of Whalton Village but is now dead. Queen was blue (born 2020).

Colours on hives represent drawing pins coloured for the birth date of the queen: White: 1, 6; Yellow: 2, 7; Red: 3, 8; Green: 4, 9; Blue: 5, 0.

Fig. 32. Appearance of Home Apiary from the South. Colours on the hives represent drawing pins coloured for the birth date of the queen: White: 1,6; Yellow: 2, 7; Red: 3, 8; Green: 4, 9; Blue: 5, 9.

98

```
18 May 2024,        (Warm and dry after a week of rain)


10 Whalton West GD1                        Saw Red queen
        6 fr. brood. Good black bees
        Bees calm          Breed from her
    X       Needs new super

Interpretation:
Hive Number 10 containing grand-daughter No. 1 of colony, caught as a
swarm at the west end of Whalton village, examined 18 May, 2024. She
emerged in 2023 (a red marker paint year), is still in residence and now
laying in 6 frames of brood and/or eggs. The bees look like natives and on
this first dry, warm day after a week of rain in May are calm. Her first super
is nearly full and she should be considered as breeding stock this season.
```

Fig. 33. Example of Field Record Book entry.

Social and Legal Issues

Honey bee conservation has to be a communal matter over which there is general agreement. Surprisingly, there is no legal obstacle that I know of that one has to overcome to declare a region a reserve, you just have to ensure that most of those with an interest in the project, especially local people, are in agreement. On the assumption there are no bee breeders or honey farmers with rather different objectives and have hives in the designated area, that usually means all the active amateur beekeepers. It also means relevant landowners and landlords, because apiary attendance requires occasional free access for one or a small number of beekeepers and vehicles.

Farmers often welcome semi-permanent hives on their land, as it can enhance their professional profile and possibly facilitate qualification for Higher Level Stewardship status, which can confer qualification for Government grants. In addition, most farmers and other country dwellers recognise the important contribution honey bees make to the natural environment. By pollinating the flowers the bees like, perhaps for the nectar they yield, bees "do their own gardening" as well as that of human gardeners and wildflower conservationists, so the wildflower profile in the area improves and benefits other aspects of wildlife as well as human activities.

Those concerned about the conservation of competitive pollinators, such as rare bumblebees and solitary bees have a valid complaint when beekeepers move several hives into an area at flowering time, just to catch a harvest from the wild flowers rare insect species depend on. But I think a small number of hives permanently sited in a conservation area (with the permission of the relevant custodians) should be acceptable. Rather than being considered competitive foragers, hive bees can be seen as collaborative pollinators, with the possible capacity to enhance the lot of all pollinator species that might use the same flowers, including rare insects using specialist plants with restricted times of flowering. It should be remembered also that our native honey bees have relatively short tongues and are unable to take nectar from some deep flowers such as those of red clover, that are eagerly visited by other bees.

Ideally the whole community should welcome small scale permanent apiaries of native honey bees and a happy acceptance can be encouraged by beekeepers rewarding landowners and others in small ways. This means keeping a low profile most of the time, respecting the lifestyles and work of country people and sharing an interest in the work they do. It is customary to reward landowners on whose land hives are sited with a few jars of honey. I suggest one jar per hive for a short visit, more for a permanent placing. The honey should preferably be from these hives on their land.

School children can be given trophies of pieces of wild comb or mouse nests found in hives and beekeepers who are able to speak in front of an audience can volunteer lectures to school classes and adults in the local community. In the area where I operate, we have encouraged a landowner to rebuild a collapsed wall of bee boles and provided literature and photographs to maintain authenticity.

Hive building classes for children with a parent, or just the parent, can be run at weekends and competitive classes at local agricultural shows, for honey, honey cookery, artwork, essays and poems can be introduced to encourage awareness and interest in bees and conservation in general. Stalls selling honey, with honey tasting and an observation hive with a marked queen and a few marked drones are always of interest and for such events third party insurance cover is available for BBKA members as a feature of membership. That insurance also protects beekeepers in the practice of their following.

In the context of personal safety adders sometimes find shelter beneath beehives on the moors. It is wise to move a stick about under every hive before you start working on them.

Finance can be a problem in setting up a reserve, particularly for purchase of hives, fencing to keep out cattle and sheep, and for making up access roads. But if local conservation bodies or the parish council see your initiative as complementary to theirs, financial help may become available. Some industrial concerns, such as wind farms, make grants available to worthy concerns, association with which can improve their image.

Publicity for a reserve can be beneficial, especially if it discourages beekeepers from importing foreign or low-quality bees. But there is the downside that it advertises the existence of high-quality stock, which is often installed in places from where it could easily be stolen. It is a good plan therefore to keep some good breeding stock at one or more secret locations.

Camouflage

A white bee suit can be seen a mile away, so attendants on hives should wear suits of camouflage material or subdued khaki colour. These days hive thieves use "helicopter drones" to locate hidden hives, so it is best if those hives also are camouflaged. Simple guidance for camouflage includes the keywords *shape*, *shadow*, and *shine*.

Shape: Paint hives with irregular patches of natural colours. In World War II, gasometers and large buildings were painted in browns and greens by three painters working independently, each from a different corner. This prevented patterns appearing which can be easy to detect.

Break up hive outlines and disguise hive roofs with turfs, logs or rocks.

Shadow: In bright sunlight shadows may reveal carefully camouflaged or uniformly spaced hives.

Shine: Hide shiny roofs by painting or sticking on artificial turf such as is used by greengrocers in their stall displays or cover the metal with roofing felt.

Camouflage nets are ideal and save a lot of work but need to be removed and replaced at every inspection. They can sometimes be purchased from Army Surplus stores, or you could make your own. (See Figure 34).

Fig. 34. A camouflaged conservation apiary. There are six hives in this view.

Chapter 9

EVOLUTION OF THE BRITISH *MELLIFERA* GENOME

To gain perspective on the genome of our native honeybees, it can be informative to consider its possible evolution with respect to recognised human influence and beekeeping practices.

Before the arrival of the Romans honey hunting was carried out probably with no attempt to keep bees in captivity. Eva Crane (1983) hints at the possible use of tree hives in Britain, as in forest beekeeping in Poland, Russia and Germany, but there is no actual evidence it took place here. It is unlikely that bee genomes underwent anything other than natural evolution in prehistoric times.

A handwritten order for *"duo lini mellari"*, two honey cloths, found at Vindolanda, beside Hadrian's Wall, shows that honey was refined there by filtering through cloth, and the discovery of a statuette of Priapus, the Roman god of fertility, suggests the possible presence of an apiary, as it was the custom for Roman beekeepers to invoke his help to protect their bees. That was from the second to the early fifth century AD.

The Roman author, Columella, a near-contemporary of Virgil, listed nine materials, including timber and straw, that were incorporated into the structure of beehives (Crane, 1983). If the Vindolanda beekeeper was from Italy he might have favoured horizontal earthenware pipes, but the horizontal aspect would have been quite unsuitable for our climate. Most of the troops on Hadrian's Wall were from what are now Germany and Belgium, and they would perhaps have been more familiar with vertical log hives, perhaps topped with a slab of stone, as still in use in the Cevenne hills of France. Columella advised stacking hives three deep on stone benches, three feet high and the same distance from front to back, but log hives would probably be too cumbersome for that style of management.

Depending on the abundance of forage, assembling many hives in one place puts pressure on the bees to forage further, and beekeepers of course would favour the most productive colonies. So there could have been what some modern beekeepers might see as "improvement" taking place then.

Saint David of Wales was born in AD500 and is recorded as ordering three of his followers to take some specific goods by boat to a new religious house in Ireland. They were a shoe full of barley, a shoe full of wheat and "a bell of bees", meaning, presumably a skep of some sort. Wicker skeps plastered with cow dung preceded those of straw, but a coiled-straw skep from Viking times and dating from around 1000AD was dug up at Coppergate in York, showing that they were in use then.

Managing colonies in skeps used to involve killing the strongest and weakest for their honey and wax, propagating only from those of intermediate size. Colonies would therefore have been bred for moderate swarming and by modern standards, low prolificacy.

The first effective movable frame hive was pioneered in the mid-1800s, by the American Rev. Lorenzo Langstroth and much of the related equipment we still use was invented soon after, including the smoker, queen excluder and embossed wax foundation. This enabled beekeepers to control swarming and to collect honey without harming the bees. Feeders enabled feeding with sugar syrup and as stated elsewhere, the quantities of white sugar beekeepers now and probably also then, fed to their colonies are well within the theoretical lethally toxic range. That practice must have had major effects on the digestive systems and gut biomes of our bees, probably increasing their vulnerability to disorders that would otherwise be innocuous and in the present time encouraging use of antibiotics and a variety of toxic chemicals. However, moveable frame hives have enormously enhanced our knowledge of honey bee behaviour and colony dynamics.

In the first decade of the 20th century, the communicable Isle of Wight Disease developed, which had a devastating effect on Britain's honey bees, killing an estimated 90% of them (see Chapter 7). This must have eliminated a vast amount of the ancestral genetic variation of the native subspecies and has probably had a major influence on the subsequent health and local adaptation of British bees.

Many foreign colonies and queens were imported to replace stocks lost to IOWD and Brother Adam commenced production of the IOWD-resistant Buckfast strain. For his time that was a forward looking and apparently wise thing to do, but is now seen as a mixed blessing, as its hybrids with native *A. m. mellifera* can be frighteningly aggressive (see Chapter 5).

Another incentive for importation was the requirement for pollinators of the newly introduced Red Clover, *Trifolium pratensis* indigenous to southern Europe and sown with rye grass to promote hay production. Its flowers are however too deep to attract native honey bees, so the long-tongued *A. m ligustica* and *A. m. carnica*, were brought in for that duty. Regrettably this also caused major introgression of foreign DNA into our native gene pool. Also, maintenance of these Mediterranean bees here, where mean summer temperatures are at least 5°C cooler (see Chapter 3), required adoption of a Southern style of bee management that has now become widely adopted into British practice, to the detriment of northern natives. For example, it includes negating natural selection, by insulating hives in winter and feeding profusely with sugar syrup.

During the 1939-45 World War, sugar was strictly rationed, but made available to beekeepers as bee food, dyed green to discourage human use. Its appearance in the supers dispelled the formerly accepted belief that it was safe to feed sugar with supers on the hive, because fed sugar does not go into the supers!

In the wake of IOWD, French and Dutch *A. m. mellifera* were also imported in massive numbers, which may have boosted native *A. m. mellifera* stocks, but has contributed to confusion over racial identity.

Introduction of novel flowering crops on a large scale, particularly the nectar-rich Oil Seed Rape, demanded adjustment by the bees of the timing of brood nest development so that colonies can take full advantage of its seasonal bounty. In heather areas brood nests are naturally timed to match its flowering in August and September and the wide availability of motorised transport has promoted use of such distant seasonal forage sites. Colonies in the Northumberland coastal plain and Tyne valley therefore now tend to reach a first broodnest peak in May-June coinciding with the OSR and a second in August in time for the heather. The interval contributes to a "July gap" when queens often go off lay.

Varroa destructor entered the country, in the 1990s, probably on secretly imported bees and it, or the viruses it carries, is recognised world-wide as the biggest killer of honey bees. It has killed very many British colonies and infected others, particularly with *deformed wing disease virus (dwd)*, the effects of which on bees can be recognised by eye. Widespread regular anti-varroa medication followed with partial success and its use contributes to additional honey bee deaths. Anti-varroa medication has also virtually eliminated the commensal *Braula coeca*, the possible consequences of which are as yet unquantified.

Native *A. m. mellifera* colonies were discovered at various remote sites in Britain and development of wing morphometry enabled their identification cheaply and with simple equipment.

Although British native bees had not previously come into contact with varroa mites, many have the inherited capacity to respond appropriately to varroa infestation, including as outlined in the Appendix 2.

DNA analysis was introduced from the 1990's, confirming subspecies' tentative identifications by wing morphometry and stimulation of a reawakened interest in native honey bee conservation, with establishment of conservation initiatives.

Disease Resistance

Chalkbrood

Honey bee diseases of all kinds are a rarity in Northumberland, except for chalkbrood and it is generally advised that the best cure for this is requeening. Cooper (1986) suggested that fairly mild outbreaks occurring after very cold nights may be the result of fluctuating broodnest temperatures. This is a characteristic of *A. m. m.* But some strains are such good house cleaners that they prevent its building up and eject diseased larvae and pupae before they can produce infective spores.

Cooper believed that susceptibility is undoubtedly inherited and inbreeding susceptible stocks seems to worsen the tendency. Stocks very susceptible to chalkbrood should therefore not be used as breeding material.

Nosema

Nosema apis is a fungal protozoan that causes dysentery. It can be eliminated in a couple of seasons by good housekeeping strains and by bees that practise comb varnishing or stripping. Excitable bees prone to excrete in the hive during manipulations or when shut in tend to be susceptible.

Foul brood

Two types of foul brood occur in Britain, American (AFB) and European (EFB). An easy way to remember a major distinction between them is that **E**uropean occurs **E**arly, before brood capping has occurred, **A**merican occurs **A**fter the brood has been capped.

There are several types of inherited resistance to AFB, but British regulations dictate that all cases must be reported and destroyed under supervision. EFB, known on the Continent as "sour brood", is less serious, but it is wise to assume all strains of honey bee are susceptible. EFB responds to antibiotic treatment, but like AFB, it is also a notifiable disease.

Wax moth

Wax moths tend to infest stored comb and can be a nuisance by spoiling the appearance of cut comb by burrowing under the cappings.

Some bee strains are tolerant of attacks by small wax moths and keeping colonies strong and in good hives can also help eliminate them. Empty comb is ignored by the moths if hung in the open air, as in an airy shed, which is useful to know with regard to winter comb storage.

Greater wax moth gets spread by fishermen emptying their bait tins at the end of the expedition, as the larvae are sold as fishing bait.

Acarine disease

Acarine disease is caused by a mite infestation of the tracheae or breathing tubes and used to be widespread in Britain. According to Cooper, most stocks are moderately resistant to acarine, but a few are highly susceptible, some of which are the outcome of crossings between the moderately susceptible.

Acarine mites are very small compared to Varroa mites and are probably vectors for chronic bee paralysis virus (CBPV), which was the probable cause of the devastating epidemic known as "Isle of Wight Disease".

Overview

Evolutionary forces with respect to honey bees since prehistory can be considered under two main headings, those of natural origin and those due directly to human influence. The former includes climatic extremes and naturally occurring diseases such as Isle of Wight Disease and Varroa, and latterly the beginning effects of global warming, whatever its causes. If left un-countered, natural selection should have taken place with respect to some of these, theoretically leaving surviving stocks with increased resistance against them.

Among humanly caused influences, introduction of new crops nudged our bees into modifying their seasonal breeding patterns, as with regard to oilseed rape in my area. Red clover was introduced to enrich the nitrate content of the soil and so facilitate the growth of grasses good for cattle to graze on. But because red clover's nectaries are beyond the reach of our short-tonged natives, bees of

Mediterranean origin with longer tongues were brought in to help and this led to the accidental incorporation of their foreign DNA into our native genomes and their widespread corruption.

The assembly of colonies in apiaries and other aspects of management have forced our bees to work harder for their harvest, to human rather than apiarian benefit, while our early use of skeps promoted their naturally lower tendency to swarm. Stealing of honey by commercial bee farmers, followed by compensatory feeding with sugar syrup, must also have had major effects on their gut biomes and probably increased susceptibility to some diseases. This has necessitated the highly abnormal practice of dosing with massive quantities of antibiotics, with all its attendant faults.

The need for predictable honey yields encourages commercial beekeepers to cosset their stocks when times are hard. However, by keeping weaklings alive there are temporary financial benefits, but longer-term detriment to the survival fitness of the stock. Indeed, the recent history of beekeeping has been overwhelmingly one of negating natural selection, which as I have stated repeatedly is in the long run a bad idea and for which we are now paying the price. On the good side for the conservationist, innovations in farming practice tend to be confined to traditional agricultural areas, like coastal plains and river valleys, less so to more inaccessible regions, in some of which the old bees still survive. These are the ones that now need our attention, which with intelligent management could be built up and come to play a more prominent part in rural beekeeping, without importation of foreign bees that make unreliable or false promises to their purchasers.

Chapter 10

CONCLUSIONS

So where are we now?

On a World scale, the prospects for the honey bee look grim! But that is perhaps because the picture is sketched from the most easily available data – that applying to large commercial concerns. Through their greed and ignorance, bee breeders and honey farmers have recklessly destroyed countless multitudes of honey bee colonies, as others recklessly all but exterminated the American Bison, thoughtlessly extinguished the American passenger pigeon and unconsciously eliminated so many other wild species that somehow got in the way. Crossbreeding of honey bees has hopelessly jumbled the ancient naturally evolved genomes, while the breeder queen strategy has consigned the inestimable treasure of the majority of its natural variation to the rubbish bins of history.

Slowly, over 50 million years, mutation of the bees' DNA and natural selection created those protective genomes, but their destruction mainly in commercial stocks has happened in just decades. Those life-saving natural assemblies of favourable alleles will never be reassembled, and the protection they conferred has had to be replaced with chemicals and antibiotics. There seems to be no way back along that route.

On the other hand, native honey bees like mine in Northumberland still exist in many obscure hideaways. I have seen wonderful local bees not only in England and Scotland, but also in Finland, Sweden, Denmark, Poland, Switzerland, France, Ireland and Wales. And I know of others in Belgium, Norway, Russia, Ukraine and Belarus. Representatives of the ancient races are still there, in the care of sensible and conscientious handlers. But by their nature they are difficult to propagate, as their natural tendency is to do everything on a small scale and conserve resources. Nevertheless, the honey my bees produce, without my trying very hard, is sufficient to pay half our weekly food bill and it is wonderful honey.

Colonies need a community

Beekeepers are well aware that individual honey bees cannot exist alone, they all need to be in a community. And in some senses, in-hive communities can also be thought of as "individuals", with their component castes and occupational specialists performing in harmony, as the organs in our own bodies. But honey bee colonies also cannot exist alone and actually have an existential requirement for others. At some stage every colony needs to reproduce and then the need for several mating partners arises.

There is something of a parallel in the requirements of flamingos to be in a group before they get in the mood to mate, a need that can be satisfied in small captive flocks by mirrors placed among or around them at mating time. But a more dramatic example is supplied by the American passenger pigeon.

John James Audubon, one of the most accomplished bird artists we have known, died in 1851. In his day, an estimated 3 billion migrating passenger pigeons darkened the American skies for days on end, only to be slaughtered by the million for human food. They were gregarious and communal and for courtship and mating required their gathering in seemingly unassailably large numbers. In their colonies, some trees had over 100 nests. But they were in decline; from 1870 that decline became precipitous and in September 1914 the passenger pigeon was officially declared extinct. In Wisconsin's Wyalusing State Park is a monument that declares: *"This species became extinct through the avarice and thoughtlessness of man."* It is already dangerously close, but we must never let that happen to the honey bee! We must respect the requirement of honey bee colonies for numerous other colonies within reach. That is an existential need! For their very survival, whole honey bee colonies need community.

Future-proofing

For his day, Br Adam had sensible aims, but his main strategy was cross- breeding and that now does not seem a good idea. However, we now know the secret of native bees' remarkable persistence for those 50 million years, the Koh-i-Noor of its genetic crown. This is the *complementary sex determiner, csd* gene, with its many sparkling allelic facets.

You may remember FB Kraus' statement quoted in Chapter 4, that the ultimate limiting factor governing the survival of a honey bee population is the number of its *csd* alleles. These are what I mean by its sparkling facets. They must continue to sparkle in great variety if honey bee populations are to survive.

In general I am not in favour of genetic engineering, but one of the latest techniques of molecular biology may have something to offer, in the re-creation of ancient, or new creation of novel *csd* polymorphisms that might yet rescue exhausted commercial stocks. This is "***CRISPR gene editing***", named from "**C**lusters of **R**egularly **I**nterspersed **S**hort **P**alindromic **R**epeats" and pronounced "crisper". This is a technique of genetic engineering by which the genomes of live organisms can apparently be safely modified. It is derived from a natural bacterial system and uses a synthetic "guide RNA" molecule, that is synthesised to define the DNA sequence selected for modification. The existing gene is then removed and a replacement one spliced in. Replacement *csd* alleles could be taken from other races of bee or synthesised anew. If managed properly, this could perhaps restore variation of the *csd* locus and reduce diploid drone production, with its devastating effects on fertility.

Another idea that comes out of this writing is the identification of ***octanoic,*** or ***caprylic acid*** as the potent natural repellent that discourages varroa mites from entering queen cells (Chapter 7). If I was troubled by varroa mites, I would be experimenting with that now.

Management priorities, Summary

In brief, then, my recommendation for genetic priorities for native honeybee conservation are:

1. Establish and maintain large populations of bees that match the wing morphometric characters of native *A. m. mellifera* and have no body colour other than grey, dark brown or black (see Chapter 3). Try to do this in areas where the vegetation has not been "improved". It is wild flowers they really need.

2. Do not destroy surplus queen cells in colonies with good morphometry, use them to expand your stocks.

3. Eliminate colonies that do not meet these standards, or destroy undesirable drone brood and re-queen.

4. Identify queen lines within the population and breed so as to ensure all native lines survive and genetic variation is maintained without introgression from other races. Do not import foreign queens!

5. Maintain or increase *csd* polymorphism by checking the brood viability with the 10 x 10 brood cell template (See Fig. 21) and propagating drones to ensure it does not fall below 75% (see Chapter 4). Otherwise unite colonies as necessary.

6. Aim to reinstate natural selection rather than artificial. It was natural selection that created the native honey bee and which may yet reinstate it.

7. Adopt the habit of taking your hives to better forage rather than feeding. Especially reduce administration of white sugar syrup or fondant except in emergencies.

8. Insulate hives in summer, but not winter, and adopt "the Scottish method for overwintering" in natural clusters (see Chapter 3).

Mellifera's dynasty

In Prospero's words in Shakespeare's "Tempest", "*Our revels now are ended.*" And with that thought I am reminded that when I taught the Beginners Class in Beekeeping at Kirkley Hall Agricultural College, I would finish each two-hour evening lecture with an appropriate bee-related poem. For my hundredth and final lecture I thought it would be suitable to assemble a new one out of Shakespeare's own words, borrowed from his many sonnets in praise of a mysterious "Dark Lady". I tried hard, but it defeated me. So instead, I have composed a new sonnet, on the same theme. This I dedicate to my father, Leonard Pritchard, who always seemed to have a poem to quote and to Mr John Morris, my English Teacher in the mid 1950's, at Caerffili Boys' Grammar-Technical School, Caerphilly, who introduced me to Shakespeare and tried unsuccessfully to persuade me to write poetry. He also introduced me to my favourite poem, John Keats's wonderful "*Ode to a Nightingale*". This then is my attempt at a sonnet in praise of the most important Dark Lady, in the style of William Shakespeare, in which that nightingale also makes an appearance.

To a Dark Lady

Thy dusky sheen doth hold me in its thrall,
Ligustica and Buckfast I eschew,
Caucasica and Carnic' like do pall
And only thy dark beauty will me do.

Thy plunging cubit index plunges lower,
Thy shift is, rightly, shifted to the left.
Thy mandibles un-man mighty Varroa!'
And should they still, we'd all be left bereft.

Unlike Keats' darkling songster of the trees,
The coming generations tread thee down,
Man's greed imports a host of lesser queens
That would usurp Mellif'ra's rightful crown.

So dance with me, sleek queen with sheen ceramic,
Let's keep Mellifera's dynasty.................... melanic!

Dorian Pritchard

APPENDIX 1

The F2 generation derived from one "breeder queen"

OVA GENOTYPES

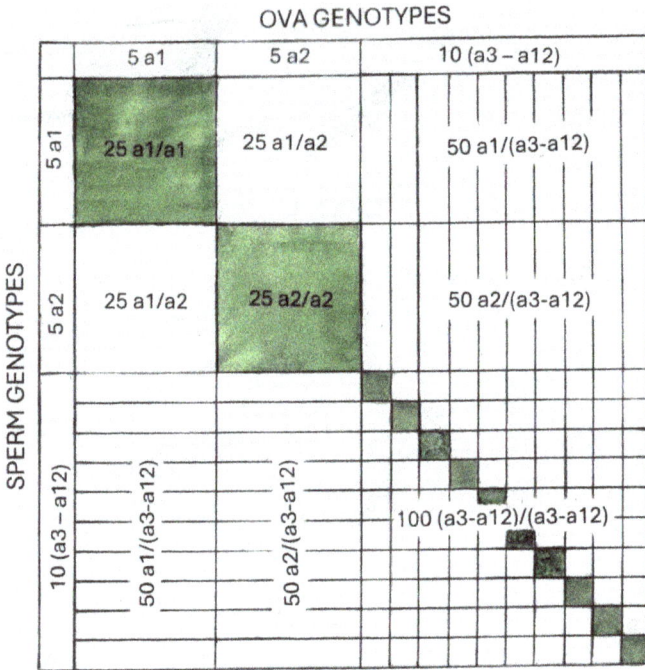

Fig 35

This diagram illustrates the outcome of the breeder queen exercise described on pages 52-54, after selection and breeding from the best of 10 queens, with no outside influences.

This representation reveals the probability of alternative outcomes of the random matings of 10 daughter queens of an *a1/a2* breeder queen and related drones in a closed community, as described on pages 52 - 54. Alternative genotypes of their ova, *a1, a2, a3* etc.to *a12* are indicated along the top and sperm genotypes vertically on the left. The outcomes of these matings result in the F2 generation represented within the square as 400 possible individuals.

The putative *csd* homozygotes are marked in green. These would develop as diploid drones and be destroyed by the house bees. They would number 25 *a1/a1* homozygotes, 25 *a2/a2* homozygotes and a total of 10 homozygotes *a3/a3; a4/a4, etc.* The total number of alternative outcomes is 400 and the total number of homozygotes: 25 +25+10 = 60, representing 60/400 or15% of the total.

It can be seen that this procedure constitutes inbreeding and would lead to a fall in fertility in the apiary, of around 15% in the F2 generation. If inbreeding were to continue within the same group without introduction of extraneous genetic material, a similar level of infertility would be expected to occur.

APPENDIX 2

How Northern native British bees, *A. m. mellifera,* overcome varroa mites.

The Varroa mite came to Northumberland in 2000 AD, but we had been expecting it for years. Two years earlier, at a SICAMM conference in Sweden, I publicly expressed my eagerness for my bees to acquire a varroa infestation, because, as a geneticist, I considered the mite a worthy opponent.

When it came, the usual people panicked, and I lost a couple of colonies. One already had an infestation of ants and I was interested to see how that would work out, so left them alone. It was unfavourably, in fact, as that colony died. However, all my other colonies seemed to shrug off the mite.

My problem in elucidating the nature of any presumed "infestation" and its strengths and weaknesses, was that I could find no mites, dead or alive, to examine. I had been in frequent communication with John Dews, who had transported native stocks from Fountains Abbey to Whitby, and was trying to breed for increased levels of damage to the mites that fell through the mesh of his varroa floor. John was aware of a German beekeeper named Lois Wallner who discovered or deduced that if 60% or more of fallen mites were seriously damaged, that mite population would be unsustainable (Wallner, 1990, 1991). As I had no microscope, he offered to examine my fallen mites for me. However, I hardly ever saw a mite and it took me two years to find six. He examined these and told me that 4 of those showed evidence as if of being bitten by a bee. This was more than 60%, but with such a small sample, that deduction could not be relied on. I had no other evidence of mite infestation and the bees continued to thrive, although varroa was in my neighbours' hives all around me.

Then in 2010, one colony, Jarrow Black 5 (JB5), developed a heavy infestation. This colony was unusual in that it had a large proportion of orange-banded workers among the dark and this presumed genetic contamination, I think, enabled the varroa to gain a foothold. Though nervous as to the outcome, I was delighted and could now set about my investigation.

It is a general principle in science that if you want to deduce how something works, you first study it without interfering. I took further inspiration from the Zulu general commanding his forces at the Battle of Rorke's Drift, as re-enacted in the epic film "Zulu". Before committing his main force to battle, he sat in a chair on the hill and watched how the British infantrymen, holed up in an isolated farmstead behaved when challenged. Armed with assegais and cowhide shields, several scores of Zulu warriors charged the hastily fortified buildings, but stopped close to the limit of their musket range and danced in a threatening way, while the soldiers did their best to shoot as many as they could. The old general just sat back and appeared to be doing nothing. In reality he was counting the British guns, assessing how long an infantryman took to reload and how accurate was his aim under extreme stress! I did similarly with my infested colony: sat back and assessed the firepower of both sides. I couldn't help wondering how those countrymen of mine would have responded to an overwhelming infestation of varroa mites!

These bees were naïve where varroa was concerned, and the mites had another advantage in having a great capacity for propagation. Being tiny, they could hide in small crevices, including inside capped brood cells. But they had the weakness that adult females needed open brood to get into just prior to its being capped at 8-9 days, in order to lay their own eggs to make up losses, or increase their numbers.

Bees I recognised also can reproduce very fast but require considerable broodnest space to do so and in this case, I found that brood space diminished more rapidly than I had expected. The bees also had the freedom to abandon the hive and if in so doing they should take away the queen, that denied the mites the open, 8-day brood and their chance to reproduce, as happened on this occasion.

The bees also had a secret weapon in an unexpected capacity, indeed eagerness, to disable and kill mites, by biting them with their mandibles. This behaviour is a variant of inter-bee grooming, or "allogrooming" (see Pritchard, 2010). But they also had an even more secret weapon, that took me years to recognise. This was apparently being able to prevent infested worker bees emerging from their brood comb, so effectively leaving their mite-infested babies "buried under the floorboards", together with the mites feeding on them. This I later deduced was the probable explanation for the unexpectedly rapid decrease in broodnest space. I suspect the cappings over infested worker brood may have been coated with propolis, as I have suspected in other hives when brood appears to have died from chilling, but I have no proof of that proposition. That behaviour has however been reported in the Eastern honey bee, *Apis cerana cerana* (Rath, 1999).

To monitor the progress of the battle I checked the behaviour of bees at the hive entrance several times a week and the broodnest weekly. I counted and recorded the number of occupied frames of brood and after the colony had swarmed, the number of emergency queen cells. I slid a Thornes' varroa floor under the brood box and counted the mites that fell through the mesh. I did this every week, examined and photographed the mites under the microscope, then plotted all the numbers I had accumulated on graphs, so as to appreciate better what had taken place. (See Pritchard, 2012, 2015).

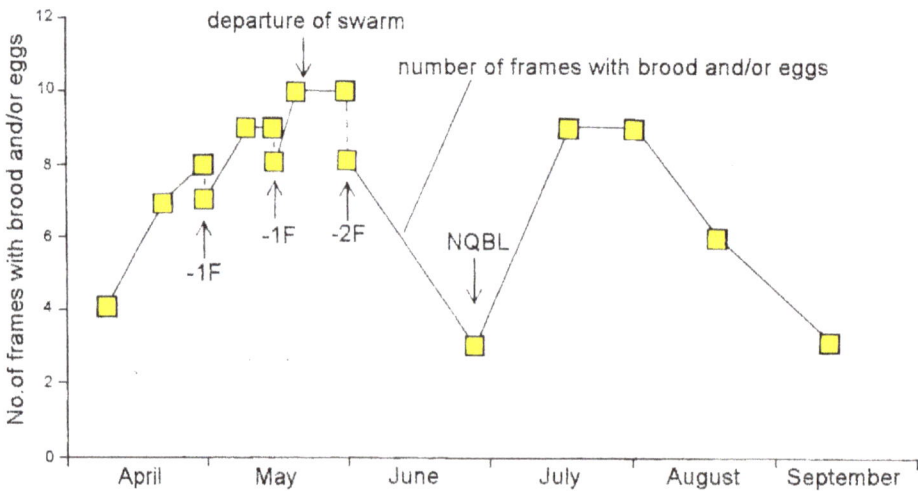

Fig. 36. Growth curve of colony JB5 broodnest. -1F and -2F indicate removal of one and two full brood frames. NQBL: New Queen Began Laying..

As mentioned above, the first thing I had noticed was that the broodbox filled up very rapidly (see Fig. 35). To slow it down I replaced two full frames of brood with empty drawn comb (see -1F in Fig. 35) and noted that the two colonies that received the two infested frames did not themselves develop infestations.

Removal of the full brood frames from the developing broodnest had little effect on it and the colony swarmed, actually while I was in the apiary. But it did so without first producing queen cells, so this was not normal swarming. I call it "pseudo-swarming". Beowulf Cooper (1986) ascribes swarming without prior construction of queen cells to overheating of the broodnest.

I failed to catch the swarm and the colony produced four emergency queen cells. I left two frames each with a queen cell in that brood box and set up two nucs from the other two. (The colonies these developed into also showed no evidence of varroa, but the queens were ginger-banded, so those colonies were later discontinued.)

Rapid build-up of the broodnest and pseudo-swarming were the first two unusual behaviours of this infestation.

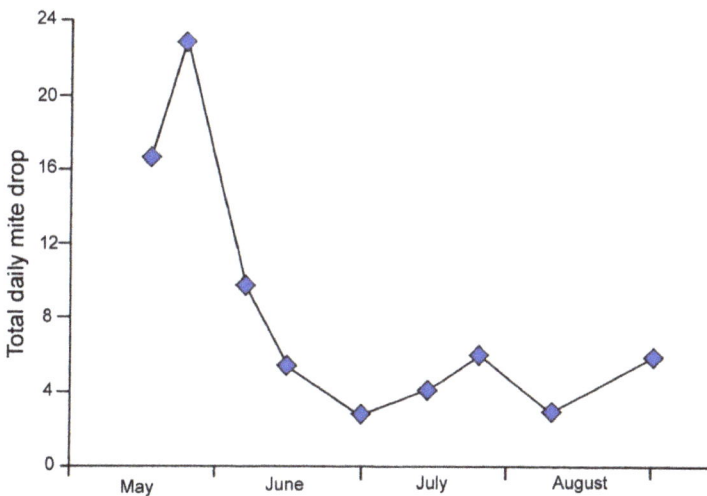

Fig. 37. Total daily mite falls in May-August. All mites were counted.

Note the high point of 23 per day in May and no indication of exponential growth except during the very earliest stage. Total mite drop ~900.

In late May, mites were dropping at 23 per day, close to four times the 6 considered indicative of an unsustainable colony (Fig. 36). The *"Managing Varroa"* booklet (FERA, 2009) also emphasises the exponential proliferation of varroa infestations, but apart from the first few weeks in May, this was not apparent in this colony. (Nor indeed have I seen it illustrated by a real example, anywhere!)

Following the late May high point, the broodnest then declined in size as the brood emerged following departure of the queen. The number of occupied brood frames should then have reduced to zero, but that did not happen, instead it got no lower than 3 or 4 frames, that at the time I thought contained live brood (Fig 35, Fig. 40, datum point 5). I now believe the brood in those 3 or 4 frames and the mites they contained never emerged and that either the developing young bees lacked the normal strength to fight their way through the capping, or the house bees prevented them getting out, perhaps by coating the capping with a thin layer of propolis (see Rath, 1999).

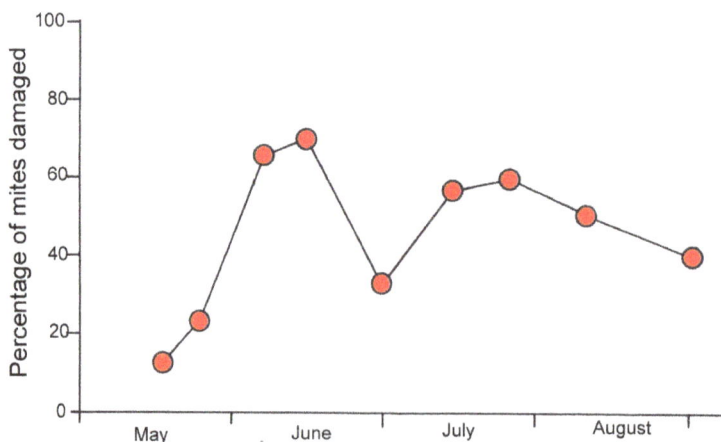

Fig. 38. Percentage of seriously injured mites that fell per day during May-August. These reached peaks of 70% in mid-June and 60% in late July.

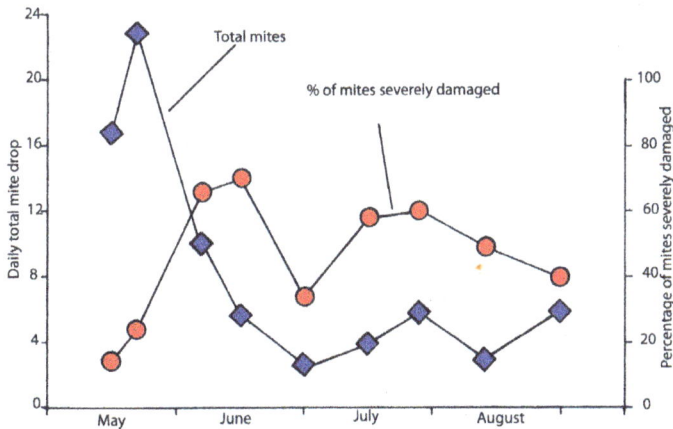

Fig. 39. Percentage of fallen severely damaged mites compared to total mite fall. Mite damage reached peaks of 70% in mid-June and 60% in late July.

While the brood nest was diminishing, following departure of the laying queen, there was a massive fall of mites and 70% of these showed severe damage, evidence of being bitten by the bees' mandibles in mid-June (Fig, 39; Pritchard, 2016). Some had pieces bitten out of their idiosomas (i.e. dorsal shields) or had one or more legs bitten off and some had both kinds of injury (Fig. 39).

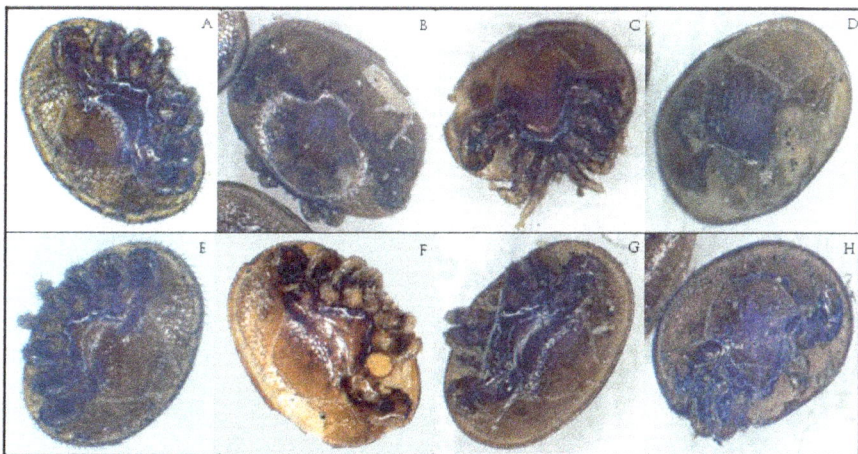

Fig. 40. Damaged mites collected from the hive under-floor. A, E: intact mites; B, C, E, F: mites showing damage to the idiosoma and severed limbs; D, H: mites that have lost all appendages.

This high incidence of injuries was found to coincide in time with the interval in brood capping at mid- to late June (Fig. 40). There was a secondary peak of 60% injuries in late July, when the total mite drop was only 6 (Fig. 36).

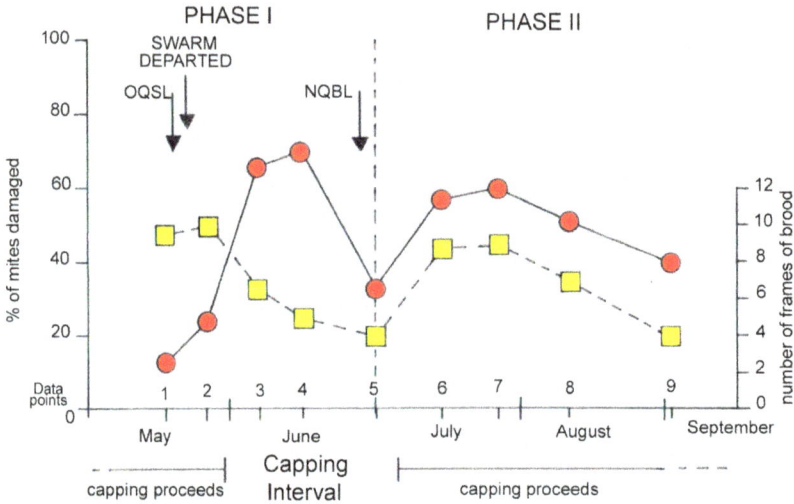

Fig. 41. Percentage of damaged mites compared to growth phases of the colony and brood capping activity. OQSL: Old Queen Stopped Laying. NQBL: New Queen Began Laying. Note that the 70% peak of mite damage coincides with the interval in brood capping and the 60% with the second growth phase of the colony.

Major physical injury of mites indicates aggressive grooming having taken place (Pritchard, 2016) and is the third unusual behaviour of these bees. The groomers taking advantage of the post-swarming brood break to attack mites constitutes a fourth aspect of the bees' defence strategy.

The larvae that emerged from eggs laid by the new queen which emerged from an emergency queen cell were capped in June, offering new hiding places for adult mites and then the mite drop decreased. But as the brood nest built up and the new cohort of young bees, including potential allogroomers, joined the hive community, the percentage of mites falling increased to peak at 60% (Figs 37, 38), but dropped again as the queen's laying rate decreased.

Changes in grooming activity may relate to changes in recruitment of young bees into the allogrooming work force, which occurs at 7 days after their emergence from the egg (Pritchard, 2016). This would seem to be regulated by the number of mites in the nest.

The proportion of damaged fallen mites then decreased and by September, they were dropping at only 8 per day, very low for this stage of an infestation (Fig. 39; FERA, 2009).

The following Spring no mites were seen in that hive, and this was also the case with the four nucs set up, i.e. neither colony JB5, nor any of the four hives that received frames potentially containing mite-infested brood developed infestations. This supports the conclusion that very few, if any of the brood cells in the translocated frames actually contained *viable* varroa mites and neither mites nor their bees emerged from them.

Overview

Throughout the season I used no chemicals and apart from the frame replacements described, made no manipulations that might disrupt the infestation. Nevertheless, the hybrid JB5 bees wiped out virtually all their mites and the colony progressed as normal. The following spring there was no obvious sign anything unusual had happened.

Although the exercise gave me considerable insight into the strategies and "fire power" of both sides, I felt no need to intervene. I know these bees had never met varroa before, but they applied their normal behaviours in such a way that they negated the advances made by the mites until virtually none were left and those few were numerically too weak for the population to survive.

Now, 14 years later, I have seen no other mite infestation in any of my hives but have taken care to keep my stocks free of ginger-banded individuals, by re-queening or giving away any colonies having bees with ginger bands. I think it is probably only the native black bees that have the defensive instinct or capacity to respond in the ways I observed, while the ginger-banded bees probably lacked the ability to act defensively by biting varroa mites.

Altogether I have managed about a dozen queen lines of all-dark native or near-native Northumberland bees and have not experienced another mite infestation. In my hands, all those colonies appeared to be close to 100% resistant to varroa. Furthermore, I have several beekeeping friends in the Northumberland area who would say the same about their locally obtained bees.

REFERENCES

Adam, Br, 1968. "Isle of Wight" or Acarine Disease: its Historical and Practical Aspects. *Bee World*, **49**: 1, 6-18. DOI: 10.1080/0005772X.1968.11097180.

Adam, Frère, 1980. *A la recherche des meilleures races d'abeilles*. *Le courrier du livre*, Paris.

Adam, Frère. 1980. *Ma méthode d'apiculture. Le courrier du livre*. Paris.

Adam, Br, 1982. *Breeding the Honeybee. A Contribution to the Science of Beebreeding.* Northern Bee Books, Mytholmroyd, Hebden Bridge. ISBN 0-907908-32-2, 118pp.

Adam, Br 1983. *In Search of the Best Strains of Bees.*

Adam, Frère, 1985. *Les Croisements et l'Apiculture de demain*. Syndicat National d'Apiculture, Paris.

Adam, Br, 1987. *Beekeeping at Buckfast Abbey; with a Section on Meadmaking.* Northern Bee Books, Mytholmroyd, Hebden Bridge. ISBN 0-907908-37-3.

Adam, Br, 1987. *Breeding the Honeybee.* Northern Bee Books, Hebden Bridge, HX7 5JS (UK).

Al Lawati, H. & Bienefeld, K. 2009. Maternal age effects on embryo mortality and juvenile development of offspring in the honey bee (Hymenoptera: Apidae). *Ann. Entomol. Soc. Am.* **102** (5): 881-888, doi: 10.1603/008.102.0514

Anderson, J. 1916. The connection of *Nosema Apis* with Isle of Wight disease in hive bees. *Proc. R. phys. Soc. Edinb.* **20**:16-22.

Ball, B.V. & Bailey, L. 1997. Viruses pp 11-32. In: R.A. Morse & K. Flottum (eds). *Honey bee pests, predators and diseases.* A. I. Root Co. Medina OH.

Bailey, L. 1958. The epidemiology of the infestation of the honeybee, *Apis mellifera* L. by the mite *Acarapis woodi* Rennie and the mortality of infested bees. *Parasitology* **48**, 493-506.

Bailey, l. 1961. The natural incidence of *Acararapis woodi* (Rennie), and the winter mortality of honeybee colonies. *Bee Wld* **42 (4)**: 96-100.

Bailey, L. 1963, *"The Isle of Wight Disease": the Origin and Significance of the Myth.* Central Association of Bee-keepers, Ilford, Essex.

Bailey, L & Ball, B. V. 1991. *Honey Bee Pathology.* Harcourt Brace Jovanovich, Sidcup, U.K.

Bailey, L., Ball, B.V. & Perry, J. N. 1983. Honeybee paralysis in England and Wales. *J. Apic. Res.* **22**: 191-195.

Bailey, L. & Lee, D. C. 1959. The Effect of Infestation with *Acarapis woodi* (Rennie) on the Mortality of Honey Bees. *J. Insect Path.* **1**,15-24.

British Bee Journal 1907 Editorial: The bee-experience in the Isle of Wight. *Brit. Bee J.* **35**:221.

Brook, B. W, Sodhi, N. S. & Bradshaw, C. J. A. 2008. Synergies among extinction drivers under global change. *Trends Ecol. Evol.* **23** (8): 53-60. doi:10.1016/j.tree.2008.03.011.

Büchler, R., 1998. *Étude comparative du development de la colonie, de l'ativité du vol et du comportement de butinage chez differente souches de carnica et de mellifera.* In: German Bee Research Institutes Seminar. *Report on the 45th meeting in Bremen.* 24-26 March 1998. *Apidologie,* **29** (5) 391-476.

Büchler, R., Costa, C., Hatjina, F. Andronov, S. et al. 2014. The influence of genetic origin and its interaction with environmental effects on the survival of *Apis mellifera* L. colonies in Europe. *J. Apicultural Res.* **53** (2) 205-214.

Bullamore, G.W. 1922. *Nosema Apis* and *Acarapis (Tarsonemus) woodi* in relation to the Isle of Wight bee disease. *Parasitology* **14**, 53-62.

Büttel-Reepen, H.1906. *Apistica. Beitrage zur Systematik, Biologie, sowie zur geschtlichen und geographischen Verbeitung der Honigbiene (Apis mellifica L.) ihrer Varietäten und der übrigen Apis-Arten.* Veröff. Zool. Museum, Berlin.

Carr, S.M. 2023. www.Nature.com/scientificreports.

Caughley, G. 1994. Directions in conservation biology. *J. Animal Ecology*.**63** (2) 215-244.

Carbon Brief. 2024. *The Atlantic Ocean current that brings warm water up to Europe from the tropics has weakened by 15% since the middle of the last century, new research suggests.* https://carbonbrief.org/atlantic-conveyor-belt-has-slowed-15-per-cent-since-mid-twentieth-century/

Climate impact gulf stream, 2024. http://www.bbc.co.uk/climate/impact/gulf_stream.shtml

Cohen, S.B. 1973. *Oxford World Atlas*, Oxford University Press. Page 89

Cooper, B.A. 1986. *The Honeybees of the British Isles.* Denwood, P. Ed., British Isles Bee Breeders' Association. Codnor, Derby, UK, 158+7pp. ISBN 0-905369-06-8.

Cooper, H. M. 1906). Bee Paralysis: is the cause known? *Brit. Bee J.* **34,** 28.

Couston, R. 1972. *Principals of Practical Beekeeping.* 101pp. Scottish and Universal Newspapers Ltd., Kilmarnock.

Crane, E. 1983. *Archaeology of Beekeeping.* G. Duckworth and Co. Ltd, London. ISBN: 0 7156 1681 1.

Cushman, D. http://www.dave-cushman.net/

Delforge, B., Guerriat, H. 2012. Pollen spectra of Buckfast and dark bees' honeys in an urban environment. *Mellifica* **101**, 10-11.

Fagan, W. F. & Holmes, E. E. 2006. Quantifying the extinction vortex. https://doi.org./10.1111/j.1461-0248.2005.00845.x

FERA. 2009. *Managing varroa*. Dept for Environment, Food and Rural Affairs, 37pp.

FranceAgriMer. 2012. *Établissement National des Produits de l'Agriculture et de la Mer:* Audit économique de la filière apicole française, Septembre.

Gilpin, M. E. & Soulé M.E. 1986. Minimal viable populations: processes of species extinction. Pp.19-34 in Soulé. Ed. *Conservation biology: the science of scarcity and diversity*. Sinauer Associates, Sunderland, Miss., USA.

Goetze, G. 1964. *Die Honigbiene in natürlicher und künstlicher Zuchtauslese*. Paney, Hamburg.

Guerriat, H. 2013. *Infatigable Abeille Noire. Revue Abeille et Cie*, **153**, Feb. 2013. Pp33-36.

Hartley, D. 1975. *Food in England*. Macdonald and Jane's London.

Hońko, S., Jasiński, Z. 2002. Comparison of different honeybee races under the conditions of south-western Finland. *Journal of Apicultural Science*, **46** (20) 97-106.

Is the AMOC approaching a tipping point? 2024. https://tos.org/oceanography/article/is-the-atlantic-overturning-circulation-approaching-a-tipping-point

Köppler, K., Vorwohl, G., Koeniger, N. 2007. Comparison of pollen spectra collected by four different subspecies of the honeybee *Apis mellifera*. *Apidology* **38** (4), 341-353.

Kraus, F. B. 2005. Requirements for local population conservation and breeding, pp 87-100. In: *Beekeeping and Conserving Biodiversity of Honeybees*.

Lodesani, M & Costa, C.,Eds 2005. *Beekeeping and Conserving Biodiversity in Honeybees*. Northern Bee Books, Mytholmroyd, UK.

Malden, W. 1909. Further report on a disease of bees in the Isle of Wight. *J. Bd Agric.* **15**(11): 809-825.

Meixner, M. D., Costa, C., Kryger, P., Hatjina, F., Bouga, M., Ivanova, E. & Büchler, R. 2010. Conserving diversity and vitality for honey bee breeding. *J. Apicultural Res.* **49** (1) 85-92.

Mooney, H. 2018. *Honey Bee Genetics*. Native Irish Honey Bee Society. 18pp.

Nazzi, F. Bortolomeazzi, R. , Della Vedovay, G., Del Piccolo, F., Annoscia, D. & Milani, N., 2009. Octanoic acid confers to royal jelly varroa-repellent properties. *Naturwissenschaften* **96** (2): 309-314, doi: 10.1007/s00114-008-0470-0

Paleolog, J. 2002. The food storage efficiency and the competition abilities in the black bee tested in Poland. *Journal of Apicultural Science* **46**. 2. 55-63.

Pettis, J. S. & Wilson, W. T. 1995. Life history of the honey bee tracheal mite (Acari: Tarsonemidae) *Arthropod Biology* **89** (3):368-374.

Pritchard, D. J. 2012. Varroa resistant bees. *Bee Craft*, **94**,6.

Pritchard, D. J. 2015. Breeding varroa resistant bees. *Beekeepers' Quarterly*, **119**: 6.

Pritchard, D. J. 2016. Grooming by honey bees as a component of varroa resistant behaviour. *J. Apicultural Res.* **55** (1) 38-48. http://dx.doi.org/10.10 80/00218839.2016.1196016

Rath, W. 1999. Co-adaptation of *Apis cerana* Fabr. and *Varroa jacobsoni* Oud. *Apidologie* **30**: 97-110.

Rennie, J. 1923. Acarine disease explained. *N. Scotl. Coll. Agric. Mem.*, No. 6.

Rennie, J. & Harvey, E. J. 1919. *Nosema apis* in hive bees. *Scot J. Agric.* **2** (4): 15-30.

Rennie, J., White, P.B. & Harvey, E. J. 1921. Isle of Wight disease in hive bees. *Trans. Roy Soc. Edinb.* **52**, 737-779.

Ruttner, F. 1987. *Biogeography and Taxonomy of Honeybees.* ISBN: 9783540177814 (978-3-540117781-4) Springer.

Ruttner, F. 1988. *Breeding Techniques and Selection for Breeding of the Honeybee.* 151pp. British Isles Bee Breeders Association. Translated by Ashleigh and Eric Milner.

Ruttner, F., Milner, E. & Dews, J. 1990. *The Dark European Honey Bee Apis Mellifera Mellifera Linnaeus 1758.* British Isles Bee Breeders Association.

Silver, J. 1907. The Isle of Wight bee-disease. *Brit. Bee J.* **35**:223-224.

Tarpy, D. R & Page, R.E. Jr. 2002. *Behavioural Ecology and Sociobiology* **52** (2), 143-150. Sex determination and the evolution of polyandry in honey bees (*Apis mellifera*).

Wallner, L. 1990. Beobachtungen natürlicher Varroa-Abwehrreaktionen in meinen Bienenvolkern (Observations of natural varroa defence reactions in my bee colonies) *Imkerfreund*, **34**, 4-5.

Wallner, L. 1991.Neue Bienen Zeitung. December 28-29. (Translation by A.E.McArthur. *The Scottish Beekeeper*, June 2012: 151-2.

Weightman, C. 1961.*The Border Bees.* Ramsden Williams Publications. Great Britain.

Webster, T. C. & Delaplane, K. S. 2001. *Mites of the Honey Bee.* 280pp. USA.

Wilde, J., Siuda, M., Rykowski, D., Bratkowski, J. 2002. Flight activity of three subspecies of honeybee depending on time of day and air temperature. *5th International Conference on the Black Bee Apis mellifera mellifera*, Wierzba, Poland. Reports and Summaries. pp 91-97.

Wilde, J., Studa, M., Bratkowski, J., 2003. Pollen collection by three subspecies of honeybee *Apis mellifera* L. *Acta Biol. Univ. Daugavp* **3** (2), 101-106.

Woyke, I. 1976.Population genetic studies on sex alleles in the honeybee using the example of the Kangaroo Island Bee Sanctuary. *J. Apicultural Res.* **15** (3-4) 105-123. https:// doi.org/10.1080/00218839.1976.11099844

Zander, E. 1907. "Tierische Parasiten als Krankenheitserriger bei der Biene." *Leipziger Bienenzeitung* **24**, 164-166.

GLOSSARY

Acarapis woodi - the acarine or tracheal mite.

Acarine disease - disease caused by acarine mites *Acarapis woodi* infesting the tracheae in the head and thorax of honey bees.

acclimatisation – physiological and other adjustment to changes in local environment and especially climate.

AFB – American Foul Brood.

aggressive grooming - grooming of other bees that can cause damage or injury to external parasites.

allele - one version of a gene.

allogroomer recruitment – as worker bees develop, they pass through a series of specialist behaviours which at 7 days post-emergence includes a transient stage of grooming other bees. Some of these allogroomers remain at this stage and recruitment describes their assimilation into the work force of the colony.

allogrooming – grooming of other bees, c.f. autogrooming, meaning "grooming of self."

American Foul Brood - bacterial disease caused by *Bacillus larvae*, which remains dormant in the larval bee until its cell is capped. It then breaks out of the larva's stomach and causes lethal septicaemia.

AMOC - Atlantic Meridional Overturning Circulation, q.v..

apiary – an area set aside for keeping honey bees.

apiary vicinity mating - mating usually of *A. m. mellifera* queens close to the hive, typically at hedge-top level when conditions do not allow flight to a distance from the home hive.

Apis cerana – the Eastern honey bee.

Apis mellifera carnica – the native bee of the area bounded by the Adriatic Sea and the Carpathian Mountains.

Apis mellifera caucasica - the native bee of the area between the Black and Caspian Seas corresponding roughly to Georgia and Azerbaijan.

Apis mellifera iberica - the native bee of the Iberian Peninsula.

Apis mellifera ligustica – the native bee of Italy.

Apis mellifera mellifera – the native bee of Europe north of the Alps and Carpathians, from the west coast of Ireland to the Ural mountains in the east.

Atlantic Meridional Overturning Circulation – movement of the water of the Atlantic Ocean caused by cold, very salty and therefore heavy water released from melting ice in the far north moving south under gravity and thereby forcing warm water to move North as the Gulf Stream.

AVM – see apiary vicinity mating

back-breeding – selective breeding towards ancestral type.

bain-marie – a device for heating the water in a vessel without its direct contact with the heat source, but instead through water held in an outer vessel in contact with the heat source. Said to be named after the bath used by Marie Antoinette.

bait hive – a hive set up to catch swarms. Ideally it should have a volume of about 40 litres, be installed around 4 metres above ground level and give out an attractive odour.

ball – balling of a queen is a response by worker bees that involves crowding her very closely, sometimes resulting in her death.

BBJ - British Bee Journal.

bee bole – a shelter built into a wall to house a bee skep in windy localities.

bee improvement – breeding bees for characters advantageous to the beekeeper.

BIBBA - the Bee Improvement and Bee Breeders' Association, a British conservation association founded by Beowulf Cooper.

biodiversity – variation among living organisms.

BKA – a regional Beekeeping Association.

brace comb – wax comb usually irregular in form sometimes linking brood or honey comb to the hive wall or to other combs.

Braula coeca – a commensal insect in the beehive that feeds on food as it is being passed by nurse bees to an adult queen. It is believed to be harmless, but is susceptible to anti-varroa chemicals and now almost extinct in Britain.

breeder queen – a queen selected from among others to be the parent of the next generation.

brood - offspring of honey bees at all stages up to adult.

brood break - a temporary pause in production of brood due to suspension of laying by the queen.

brood frames – usually deep, wooden frames that hold the brood and also stores of honey and pollen in the brood nest.

brood nest, broodnest – the part of the hive where brood is raised, usually in a deep box separated from the main honey stores by a queen excluder.

brood viability - the proportion of the wax brood cells that contains live, developing bees.

brood-and-a-half - arrangement of the brood nest on both shallow frames in a shallow box and deep frames in a deep box, with no queen excluder between them.

Broodfood, brood food - also known as "bee milk"; a secretion of the paired hypopharyngeal glands of nurse bees, the main food of the larvae.

Brother Adam - for many years the leading beekeeper of Buckfast Abbey, Devon. Born Karl Kehrle

Buckfast bee - a multi-hybrid strain bred by Br Adam at Buckfast Abbey in Devon, founded on Ligurian *A.m. ligustica* and native British *A.m. mellifera* crosses between which were found to be resistant to IOWD.

C- lineage - one of five or more ancestral lineages of subspecies of the species *Apis mellifera*.

Candipolline Gold – a mixture of sucrose, glucose, sugar syrup, bee pollen, caseinate, albumen and glycerine advertised as a complete food for honey bees, manufactured in Italy.

capping interval - the period between cessation of capping of brood following cessation of laying by the queen and its recommencement following the queen's return to laying.

caprylic acid - octanoic acid, found in goats' milk. As a component of the royal jelly of queen cells it repels varroa mites.

cast - a second or subsequent swarm after the prime swarm has left the hive.

caste- a specialised group of individuals among social insects; for example, workers and queens are both genetically defined as female, but develop with different specialisations due to variant nutritional regimes (*c.f.* cast, *q.v.*)

CBPV – see Chronic Bee Paralysis Virus.

Central Association of Beekeepers – the CABK is an educational charity registered in the UK, the objective of which is to promote and further the craft of beekeeping. By organising lectures and producing publications, they aim to act as a bridge between the beekeeper and the scientist.

chalkbrood - this results from a larva eating spores of the fungus *Ascosphaera apis* that germinate and send out strands of mycelium that spread throughout its body, converting it to a hard "mummy".

Cheviot – a mountain mass reaching a height of over 2600 feet on the English-Scottish border.

chromosomal DNA - the main body of an organism's DNA that forms a significant part of all its chromosomes and carries most of the species' genetic information in coded form.

Chronic Bee Paralysis Virus - a virus which causes crawling from colonies and with the acarine mite, was responsible for IOWD.

CI – cubital index; a measure of a wing feature utilised in distinguishing subspecies of *Apis mellifera*.

C-lineage - one of five or more ancestral lineages of subspecies of the species *Apis mellifera*. The C-lineage contains *A. m. carnica* and *A .m. ligustica*.

coefficient of inbreeding - the probability that the two alleles of a gene in a homozygote are derived from the same ancestor.

coiled straw skep - a skep hive made from a rope of twisted straw sewn into the shape of a basin.

collaborative pollinators - pollinators of different species that effectively assist one another by both pollinating a particular flower species.

Colony JB5 – one of my colonies of the Jarrow Black Newcastle strain.

COLOSS - an International Association addressing the causes of honeybee losses.

colour marking of queens - the year in which a queen was born is indicated by a spot of coloured paint on the dorsal thorax. Years ending in 1 or 6 are marked White, 2 or 7 Yellow, 3 or 8 Red, 4 or 9 Green and 5 or 0,Blue. The mnemonic "Will You Rear Good Bees?" serves as a reminder of this code.

comb capping - capping of brood comb, which occurs at around 9 days. after the egg is laid in all three bee castes.

comb varnishing - lining of brood cells with propolis, which acts as a disinfectant.

competitive pollinators - pollinator species that compete by taking pollen from the same flower type. (See "collaborative pollinators".)

Complementary Sex Determiner, csd – the gene responsible for defining sex in bees; a single copy of an allele defines maleness, heterozygosity, femaleness and homozygosity a diploid drone.

conservation - preservation, protection, and management of healthy living specimens of one or more endangered species, or an environment, in order to secure their/its survival and welfare into the distant future.

conservation area - a nature reserve.

Coppergate – a street in York where the archaeological investigation of a Viking settlement was undertaken.

CRISPR gene editing - a molecular genetic method for replacing specific genes in live organisms.

cross-breeding - reproduction by mating between two lines or subspecies.

Cubital Index – a term that describes the ratio of the lengths of two regions of a specific wing vein in honey bees.

daily mite drop - the average number of varroa mites that fall to the floor of the hive within a specific time interval.

Dark bee - a popular name for the native British *A. m. mellifera*.

Darwin's Theory - now no longer only theory; the explanation of how evolution of species occurs through natural selection of the fittest individuals and genetic transfer of their beneficially adaptive characters to their offspring.

deformed wing disease - shrivelled wings of adult bees due to infection by the *dwd* virus. Usually related to varroa infestation.

Dews' plot - A graph of Cubital Index in relation to Discoidal Shift used to identify subspecies identity of a honeybee colony, a method developed by John Dews of Whitby.

diploid - having two sets of chromosomes. Female bees are all diploid.

diploid drone – a sterile, ostensibly male honeybee produced by homozygosity of the *csd* gene usually due to inbreeding. They are normally destroyed by house bees immediately after emergence from the egg.

Discoidal Shift - a term that describes the relative positions of specific wing vein junctions in a bees' wing. It is described in degrees positive or negative.

DNA analysis – DNA carries the genotype of a species, including instructions for synthesis of its proteins and guiding its anatomical and behavioural development. It notably enables investigation of the evolution and relationships of species at a level below that of the phenotype, i.e. the observable animal or plant.

drawn comb - empty honey or brood comb.

drifting - the tendency for foraging bees sometimes to return to a neighbouring hive, usually down-wind.

drone assembly - the gathering of drones at a chosen place, often in a thermal, which queens also visit in order to mate.

drone congregation - drone assembly.

drone foundation - wax foundation embossed with a large-scale hexagonal pattern to stimulate building of drone cells and drone production.

drone-layer – a queen that produces only drone offspring due to her failure to mate, so that she lays only unfertilised eggs which develop into drones. Or an old queen that has exhausted her store of sperm.

drone-producing colony – a colony the beekeeper has chosen to be the male parent of the next generation by insertion of one or two brood frames loaded with drone comb or drone foundation.

Dryas octopetala - the Mountain Aven, a cold tolerant plant with eight-petalled white flowers used as an indicator of temperature in the recent past, such as "the Younger Dryas".

DS – Discoidal Shift.

dwd - deformed wing disease, usually caused by a virus that is transmitted by the varroa mite. The presence of bees with shrivelled wings usually means that colony has a varroa infestation.

emergency queen cell – a queen cell produced by a colony to replace a queen that has died or been lost.

epigenetic - an aspect of phenotype conferred by perhaps physiological or environmental influence that modifies the outcome of straightforward gene expression.

essential oils - plant oils that vaporise readily, for example lavender. Some have therapeutic properties.

European Foul Brood – a disease caused by *Melissococcus pluton*, a very small non-spore-forming bacterium. This enters the body of a larva in the brood food and proliferates in the stomach of the larva, which generally dies of starvation before its cell is sealed.

existential challenge - a challenge so severe that it threatens the existence of the population or colony.

expanded- foam house insulation - very lightweight plastic foam sheets, covered in silver foil. It is supplied in several thicknesses which are easily cut with a knife or saw.

exponential proliferation - exponential growth sometimes occurs in bacterial or eukaryote cell cultures and is usually described in terms of doubling time, which is supposed to continue indefinitely. This rarely, if ever occurs in nature as frequently described by theoreticians.

extinction vortex - when species have gone extinct, they have tended to go through a similar sequence of events: reduction in population size, inbreeding, inbreeding depression and further reduction in population size. This ever-worsening cycle of events is likened to a swimmer or kayaker being caught in a whirlpool.

F1 generation - the "first filial generation", i.e. the sons and daughters of the mated pair.

F2 generation - the "second filial generation", i.e. the granddaughters and grandsons of the mated pair.

F2 aggression – bad temper of worker bees that occurs following crosses between bees of different temperament groups, but usually not in the F1 generation.

Fat bodies - honey bees have fat storage areas beneath their external carapace. Northern native bees obtain fat from pollen in late summer, enabling them to survive the winter with little consumption of hive stores.

feeding bees - the usual way to feed bees is with syrup made from white sugar (sucrose) and water placed in a feeder with restricted access for the bees. In winter fondant is generally preferred as it does not introduce so much water into the hive.

FERA - Food and Environment Research Agency.

feral - a species or individual that has escaped from captivity or domesticity and returned to the wild.

following – bees that aggressively accompany the beekeeper for a considerable distance after a hive inspection are said to "follow" or show "following". It is an undesirable trait that should be bred out, but may be tolerated if it discourages unwelcome intrusion into the apiary.

fondant – essentially a paste of white, finely ground sugar mixed with glucose syrup. It is favoured as a winter bee food as it can be given without introducing much water into the hive.

forage -flowers from which bees can collect nectar and pollen.

foraging – the act of collecting nectar and pollen from flowers.

forest beekeeping – a style of beekeeping used in Russia, Poland, etc. involving use of vertical or horizontal hollow log hives on the ground or suspended in trees, or by cutting cavities in living tree trunks and fitting them with doors.

foulbrood – EFB and AFB are two types of foulbrood.

Fountains Abbey – one of the largest and best preserved ruined Cistercian monasteries. Situated 3 miles SW of Ripon in N. Yorkshire, built in 1132. The discovery of a colony of feral native *A. m. mellifera* there by William Bielby in 1966 disproved claims that the subspecies was extinct in Britain.

free mating – unrestricted mating.

gene flow – the slow transfer of genes through and between populations.

gene pool – the combined genomes of a population.

genetic contamination – unwanted spread of foreign or deleterious alleles into a protected gene pool.

genetic engineering – the use of molecular genetic techniques to modify the genome of a living animal or plant.

genetic introgression – past introduction or invasion of foreign DNA into a naturally evolved genome.

genetic variation – the existence of genetic variety or polymorphism within a population. It is one of the great strengths of native populations of all species.

genetically compromised – said of a natural population that has acquired alleles that confer previously absent weaknesses.

genome – strictly all the genes carried by a single gamete, but also to denote the total genetic information held by a population.

genotype – the information content of the genome of an organism, as distinct from its physical representation or phenotype.

gentleness – calmness of a colony with disinclination to sting or act in an exaggeratedly defensive manner.

ginger-banded workers – worker bees with an orange band on the first abdominal tergite indicative of foreign ancestry, probably usually Carniolan.

global warming – the world-wide increase in atmospheric temperatures.

Great Frost – several years of very low winter temperatures, especially in 1709 when the Thames notably froze over at an estimated -12ºC.

Great Ice Age – the period between 13 million and 10,000 years ago during the Pleistocene epoch when much of the Northern hemisphere was covered by sheets of ice, though with occasional warmer interglacials.

grooming – use of the mandibles to re-adjust the body hair, akin to combing and also to remove pollen and ectoparasites, notably varroa mites.

guard bees – bees that check and if necessary, repel wasps and potential robber bees at the hive entrance.

guide RNA – ribonucleic acid with a pattern of bases matching those in the DNA close to a gene or other region of interest that enables its targeting.

gut biome - the population of bacteria and other microorganisms that occupy the alimentary canal.

habitat degradation - reduction in quality of a natural habitat, for example by agriculture or industrialization.

haemolymph - body fluid in invertebrates that performs a similar function to blood in mammals.

haploid - having only a single set of chromosomes, as in sperm ova and drone bees.

heterozygosity - the existence of two different alleles of a gene in an individual.

Higher Level Stewardship - a category of qualification of farmers for application for financial support for projects with ecological value.

hive abandonment -emigration of the whole colony from a hive.

hive economy -the balance between income and use of food in a beehive.

homozygosity - the existence of two identical alleles of a gene in an individual.

honey bee - a member of the genus *Apis*. It includes the West European species *Apis mellifera* and the eastern honey bee, *Apis cerana*.

honey cloth - a cloth for filtering honey.

honey farmer - the profession of honey producer as distinct from honey marketer and bee breeder.

honey flow - strictly nectar flow. A time when conditions are particularly favourable for plants to secrete their nectar.

honey hunting – searching for and taking honey from wild bee nests.

honey sac - the honey stomach.

hybrid - a plant or animal resulting from a cross between parents that are genetically unlike. In beekeeping, it usually refers to the outcome of crosses between different subspecies of *Apis mellifera*.

hybrid vigour – heterosis; the greater vigour in terms of growth, survival and fertility of hybrids. It is always associated with increased heterozygosity.

idiosoma - the dorsal plate of the varroa mite.

improvement - enhancement of inherited characters by selective breeding, usually in the context of the honey harvest, aggressive behaviour, or swarming.

inbreeding – The crossing of closely related individuals.

inbreeding depression - decreased vigour in terms of growth, survival or fertility following one or more generations of inbreeding. In honey bees, the most dramatic feature is numerous diploid drones.

insulation – use by the beekeeper of materials of low heat conductivity in order to prevent colonies losing their heat.

invasive species - species that tend to invade territories and overtake their native inhabitants.

IOWD - Isle of Wight Disease. A highly transmissible infective disorder that devastated British bees in the first two decades of the 20th century. It is caused by chronic bee paralysis virus, usually carried by the acarine mite.

isolation starvation - starvation of bees in the winter due to creation of a gap between the colony and its stores.

Jarrow Black bees - a strain of dark bees native to the Newcastle area.

Jenter cage - A plastic device for queen propagation. It causes queens to lay in the ends of removable plastic plugs, which can then be taken out and transferred to modified brood combs for development into queen cells.

Jenter plugs – see "Jenter cage".

John Dews - with the Rev. Eric Milner, the pioneer of the optical method of wing morphometric assessment and its graphical analysis.

July gap – equivalent to the June gap in southern Britain. A reduction in floral abundance in the North of England between June and August, when queens often cease laying.

junk food – food that temporarily satisfies hunger but does not provide the balanced diet necessary for normal health, activity and growth.

keystone species - species that play key roles in the survival of other species due to their effects on the wider environment. Examples are the beaver, the earthworm and the timber wolf.

lactic acid bacteria – beneficial bacteria so called because they control the acidity of gut contents by producing lactic acid as an end point of their carbohydrate metabolism. This makes an unfavourable environment for many pathogenic microorganisms aided by production of proteinaceous bacteriocins which they also produce.

land bridge - this was an expanse of land linking the British Isles to continental Europe following the melting of the North Sea ice sheet at the end of the Ice Age. It was flooded again about 9000 years ago.

larva – the feeding form of the young bee that precedes pupation.

laying workers - These are worker bees that develop ovaries and lay eggs, which can occur especially in the absence of a functional queen.

LD$_{50}$ - the dose of a toxic chemical that kills 50% of the rats in a test population.

lehzeni – a one-time species name for heath-dwelling northern Dark bees, distinguishing it from the forest-dwelling *sylvarum* ecotype. That distinction is no longer recognised.

Little Ice Age -a cold period in British history, extending from 1305 to 1860.

local adaptation - physiological and behavioural adjustment to local conditions.

local bee – a bee native to the locality.

M-lineage - one of five or more ancestral lineages of subspecies of the species *Apis mellifera*.

mandibles - the equivalent of jaws in insects used for manipulation of wax and propolis, grooming and fighting.

mating nuc – this is a small colony set up from a hand-full of bees and a virgin queen created to promote her chances of mating.

mean value – the arithmetic average of a set of values.

Mediterranean bees – mainly used to denote *A. m. ligustica* and *A. m. carnica*.

mellifica - Latin for" honey-maker" a one-time alternative name to *mellifera*, meaning "honey-carrier".

Mendelian reasoning - Gregor Mendel revolutionised our understanding of inheritance with his recognition that heritable information is in quanta, not fluid essences, as was previously believed. He visualised packets of information transmitted in the ova and sperm, which we now call genes. He deduced that it makes no difference whether they are transmitted by the male or the female parent and that some can be dominant to or recessive to others.

mitochondrial DNA -the mitochondria are energy generating intracellular organelles that have evolved from symbiotic bacteria. They each still retain a small chromosome based on a chain of DNA, as in the nuclear chromosomes, but these are transmitted to offspring only by reproductive females i.e. the queens.

M-lineage - one of five or more ancestral lineages of subspecies of the species *Apis mellifera*, containing the Northern subspecies *Apis mellifera mellifera*.

Mountain Aven – *Dryas octopetala*, a mat-like mountain plant that grows especially well in cool areas.

mouse guard – a strip of flexible metal perforated with holes that are larger than a worker bee but smaller than a mouse to prevent entry of mice into the beehive in winter.

moveable frame hive – a hive from which the frames can be removed for inspection or honey extraction.

native bee - from the British perspective, *Apis mellifera mellifera*.

native beekeeping - beekeeping with observance of the special requirements of native honey bees.

natural beekeeping - beekeeping in as natural a way as possible, for example with no queen excluder or swarm control.

natural selection – selection of the fittest individuals by natural forces. In Charles Darwin's theory of evolution, natural selection is the main agent of evolutionary advance.

near-native – a largely native stock, but with some (perhaps 5-10%) introgression of foreign genetic material.

nectar -the sugar-rich secretion of flowers.

nectar flow - high rate of nectar secretion by a group of plants.

negative DS -Discoidal Shift to the left in the conventional Dews' plot, as shown by *Apis mellifera mellifera.*

nine-inch hollow - a design of concrete building block ideal for use as hive stands, as a hive strap can be threaded through it.

non-prolificacy - low rate of reproduction.

North Atlantic Conveyor – the AMOC, circulatory currents in the North Atlantic that create the warm Gulf Stream.

Northern bees – the M-lineage, resident *Apis mellifera mellifera.*

Nosema - a gut infestation by the protozoan *Nosema apis* and the symptoms it causes.

nuc – a small, queen-right stock of bees.

nurse bees - newly emerged bees that feed the larvae and maintain the nursery area.

oasts - kilns for drying hops.

observation hive - a usually flat hive fitted with a glass window, through which the behaviour of its bees can be examined and studied.

octanoic acid – caprylic acid, an 8-carbon acid notably found in goats' milk and queen cell royal jelly, that is repulsive to varroa mites.

oil seed rape - a member of the Brassica family grown for the oil in its seeds and renowned among beekeepers for its voluminous secretion of nectar.

OSR – oil seed rape.

outbred - resulting from a mating between unrelated partners.

ovum/ova - an egg, eggs.

passenger pigeon – *Ectopistes migratorius* an American dove that in the 1870s was numbered in billions, but became extinct in 1914 due to its mass slaughter by man for food.

pepperpot brood – worker brood with many gaps, as if shot by a shotgun.

physiologically juvenile - able to secrete broodfood and behave like a nurse bee.

pipe – a high-pitched noise produced by queens when in their cells before emergence.

pollen - the plant equivalent of sperm, which has to be transmitted between flowers before they can bear fruit or seeds. Some species rely on insects, especially honey bees, to transfer pollen. It is an important protein and lipid food source for bees.

polymorphism – the existence of a variety of alleles of the same gene within a population.

positive DS – Discoidal Shift to the right on a Dews' plot.

Priapus - the Roman god of fertility represented as having an enormous erect penis.

probiotic - a supplementary foodstuff containing beneficial bacteria and other microorganisms that has a positive influence on good health.

propolis - a sticky, resinous material that is gathered by bees from trees, in Britain especially poplar. It has disinfectant properties and is used by the bees for lining brood cells, stopping holes and sticking wax comb to the timber hive body.

propolisation – coating with propolis.

pupa - the stage in the life cycle of a honey bee when the fully grown larva disintegrates and reassembles as an adult bee.

pygmy shrew – *Sorex minutus*, a tiny insectivorous mammal that can enter hives through the mouse guard and kill colonies

queen line - an extended family of bees descended from one queen and related through the female line.

queen substance - a mandibular gland secretion of the queen containing fatty acids that act as pheromones: 9-oxydeconoic acid is a drone and swarm attractant; 9-hydroxydeconoic acid, an inhibitor of supersedure queen cell construction and ovary formation in workers.

queenright - having a functional queen.

radial extraction - centrifugation to extract honey from combs, with the plane of the comb aligned along a radius of the rotor.

rare amino acids - the bodies of honey bees contain amino acids and proteins synthesised from them, but their relative concentrations are not necessarily similar in the pollen on which bees feed. From the perspective of the bee, some amino acids can therefore be considered rare.

rare bumblebees - bumblebees that are uncommon in the region concerned.

recessive disease - hereditary diseases caused by homozygosity of deleterious recessive alleles.

rectal catalase - an enzyme in the rectum of native Dark worker bees that helps eliminate faecal gas.

refugia – warm refuges used by honey bees during the northern glaciations of the Great Ice Age.

requeening - replacing the queen.

royal jelly - broodfood or bee milk which when given to queen larvae contains a higher concentration of sugar thought to initiate their transformation into reproductively capable queens, as distinct from sterile workers.

severe damage (to varroa mites) - this typically includes severance of legs and damage to the idiosoma due to aggressive use of the mandibles.

shot brood – pepperpot brood, worker brood that looks as if it has been hit by shot from a shotgun due to gaps caused by removal of diploid drones by the house bees.

shrew guard - similar to a mouse guard but cut from a sheet of queen excluder that allows worker bees to pass through, but not pigmy shrews.

SICAMM – *Societas Internationalis pro Conservatione Apis Melliferae Melliferae,* the International Association for Conservation of the Northern Dark Bee.

skep - a bell-shaped, open-bottomed hive usually made from a coiled rope of straw or wickerwork cloomed with mud or cowdung.

SNP analysis – analysis of the types and combinations of variant single nucleotides in DNA.

solid floor – a hive floor without a mesh window.

solitary bees – bees not closely related to honey bees that are not communal and make small nests for each offspring.

sour brood – European Foul Brood.

south-adapted – adapted to southern, i.e. a warmer climate than that of the North of England.

Southern bees – bees that originated around the Mediterranean.

Southern style – the mode of beekeeping found to be appropriate to southern rather than northern bees.

spores – the equivalent of seeds in fungi.

spring dwindling - failure of a colony to build up in spring.

stores -honey and pollen stored in the comb. Honey is stored in both the broodbox and the supers, pollen usually only in the brood box, when the hive is fitted with a queen excluder.

sucrose - ordinary white sugar. This is a disaccharide that can be split enzymically into two monosaccharides, glucose and fructose, a process sometimes referred to as "inversion".

super – an upper, usually shallow box to hold shallow honey frames.

supersedure - replacement of a substandard queen by a young daughter who initially works in support of her mother without aggression between them.

supplementary feeding – feeding usually with sugar syrup additional to normal.

sylvarum - a supposed woodland ecotype of *A. m. mellifera* now not used.

synergy - the combined effect of two or more influences being greater than the sum of those of the individual influences.

Tarsonemus woodi - the old name for *Acarapis woodi*, the acarine mite.

the cold way – brood frames fitted perpendicular to the hive entrance slot.

"the Scottish method" of over-wintering - a hive setup designed by an Edinburgh team to promote air circulation around the colony and allow the hibernating colony to hang in a cluster.

the warm way - brood frames fitted parallel to the hive entrance slot.

tomenta - the light bands across the abdominal tergites.

tongue - the proboscis.

tongue length – *A. m. mellifera* has a relatively short tongue that is inadequate for feeding from some deep flowers such as red clover.

top insulation - insulation between the crown board and roof.

tracheae - breathing tubes that allow air circulation inside the head and thorax of the honey bee.

tracheal mite – the acarine mite, *Acarapis woodi*.

tree hive - a hive carved out of a living tree trunk or positioned among the branches to secure it from brown bears.

uncapped - having had its wax cap removed.

undrawn foundation - wax foundation that has not been developed as comb.

uniting for winter - combining two or more colonies to make a sufficiently strong colony to survive the winter.

varroa - *Varroa destructor*, a parasitic mite of western honey bees, formerly considered to be *V. jacobsoni*.

varroa floor - a hive floor with a mesh panel to allow mites to fall through and be collected on a tray below.

Varroa resistance - capacity to resist varroa infestation by whatever means.

varroa resistant - describes bees that actively resist and kill varroa mites rather than merely tolerating them.

Varroa Sensitive Hygiene - a behaviour of bees involving uncapping of varroa infested cells and ejecting the mites inside along with the larva.

varroa trap - a putative means by which one colony of Northumberland bees apparently enclosed many live adult mites in a modified queen cell.

vernalization - a process in which young plants are treated with low temperatures to induce a change to an older physiological state, actually the reverse of the age change brought about in winter bees. Vernalization can be reversed by high temperatures.

Vindolanda- a Roman fort beside Hadrian's Wall.

VSH – Varroa Sensitive Hygiene, q,v.

wasp trap - a vessel that allows entry of wasps attracted to bait, but does not permit their escape.

wax moth – the Greater Wax Moth is *Galleria mellonella*, the Lesser is *Achroia grisella*. They are among the very few animal species whose larvae can feed on beeswax.

wicker skep - a primitive bell-shaped beehive of woven Willow stalks coated with mud or cowdung.

wild – said of honey bee colonies living in natural cavities, untended by humans.

wild comb - comb that has been constructed by bees without the guidance of a sheet of foundation.

wild swarm – a swarm generated by a wild colony.

wing morphometry - the measurement of insect wings by standard methods.

winter bees – young northern bees that have built up their fat stores enabling them to survive the winter and be physiologically adjusted so as to retain the capacity to produce broodfood the following spring.

winter cluster - the swarm-like congregation of a winter colony of northern bees suspended below the brood nest.

Younger Dryas - a period lasting 1200 years after the main end to the Great Ice Age, when British temperatures dropped sufficiently low that *Dryas octopetala*, normally a mountain plant, was able to grow in the British lowlands.

zygote – the diploid product of fertilisation of a haploid ovum by a haploid sperm.

INDEX

K

keystone species 14
Kirkley Hall Agricultural College 11, 117
Kraus, F. B. 30, 50, 52, 113

L

lactic acid bacteria 43
Langstroth, Lorenzo 106
laying workers 78
lehzeni 62, 63
Linnaeus 62
Little Ice Age 45
local adaptation 34, 79, 105
longevity 62

M

mandibles 77, 85, 116, 119, 122
Mediterranean bees 34, 35, 105
mellifica 62
Mendelian reasoning 64
Miller, Kyle 3, 25
mite fall 120, 122
mitochondrial DNA 12, 21, 23, 24, 27, 28, 65
M-lineage 12, 21, 63
Mountain Aven 5
mouse guard 39

N

native bee 5, 62
natural beekeeping 5
natural selection 8, 12, 14, 72, 105, 108, 109, 111, 114
near-native 11, 28, 48, 57, 82, 125
nectar 6, 31, 34, 43, 62, 68, 69, 73-79, 99, 100, 105
nectar flow 31, 73, 79
non-prolificacy 76

queen substance 74

R

rare amino acids 42
rare bumblebees 100
recessive disease 49, 83
rectal catalase 36
red clover 100, 108
refugia 5
Rennie, John 91, 92, 94
requeening 87, 106
robbing by wasps 97
Rorke's Drift 118
Rothamsted 91, 93
royal jelly 86
Ruttner, Friedrich 17, 20, 32, 33, 81, 83

S

seasonal management 38, 39
secret weapon 22, 49, 80, 118, 119
selection 8, 12, 14, 15, 21, 22, 28, 58, 72, 83, 105, 108, 109, 111, 114
severe damage (to varroa mites) 122
SICAMM 1, 2, 12, 22, 117
skep 104
smoker 65, 104
SNP analysis 22
solitary bees 100
sour brood 88, 107
Southern bees 35
Southern style 105
sperm 23, 53, 80, 81
spring dwindling 90
stores 15, 34, 35, 36, 39, 68, 71, 72, 81, 87, 96, 97, 102, 107
sucrose 6, 36, 42, 68
super 34, 38, 39, 41, 97

www.ingramcontent.com/pod-product-compliance
Lightning Source LLC
Chambersburg PA
CBHW041601260326
41914CB00011B/1347